BIOETHICS

Foundations, Applications and Future Challenges

Editors

Irene Cambra-Badii
Universitat de Vic – Universitat Central de Catalunya
Spain

Ester Busquets-Alibés
Universitat de Vic – Universitat Central de Catalunya
Spain

Núria Terribas-Sala
UVic-UCC Chair of Bioethics, Fundació Grífols
Spain

Josep-E. Baños
Universitat de Vic – Universitat Central de Catalunya
Spain

CRC Press
Taylor & Francis Group
Boca Raton London New York

CRC Press is an imprint of the
Taylor & Francis Group, an **informa** business
A SCIENCE PUBLISHERS BOOK

First edition published 2024
by CRC Press
2385 NW Executive Center Drive, Suite 320, Boca Raton FL 33431

and by CRC Press
4 Park Square, Milton Park, Abingdon, Oxon, OX14 4RN

CRC Press is an imprint of Taylor & Francis Group, LLC

Library of Congress Cataloging-in-Publication Data (applied for)

ISBN: 978-1-032-21755-0 (hbk)
ISBN: 978-1-032-21756-7 (pbk)
ISBN: 978-1-003-26988-5 (ebk)

DOI: 10.1201/9781003269885

Typeset in Palatino Linotype
by Radiant Productions

Preface

A new look for Bioethics: looking backward for looking forwards

Bioethics is a discipline that addresses issues of great relevance to human beings. For more than four decades it has generated growing interest on the part of academics, clinicians, educators, researchers, and the general public.

Already in the second half of the 20th century, the development and application of life and health sciences posed very important challenges for society. In the 21st century, great achievements, especially in the field of biomedicine, have exceeded the limits of laboratories and healthcare professionals. With techno-scientific advances, endless possibilities opened up, like the discovery of the structure of DNA and the human genome, advances in genetics, the definition of brain death, and the resulting debates on euthanasia and organ donation, assisted reproductive technologies, among others.

All these scientific advances have different connotations for humanity and made it necessary to analyze the ethical commitments in each situation. Perhaps our society has unconsciously assumed the principle that *everything that can be done technically should not always be done because it is not ethically correct*. Scientific progress cannot be made without considering the consequences that it may have in a collateral way beyond the strict objective that defines the scientific field.

Today we believe that the social implication of bioethics is evident and that professionals from all fields must have knowledge of the fundamental aspects of this discipline to form their own opinion and be able to respond in those situations that are close to them. In the same way, citizens must have a minimum knowledge of the issues that affect their lives to form criteria and a critical spirit in situations and contexts in which they must make decisions. In other words, bioethics must be part of both the basic training of professionals and citizens, in a society that takes into account the biopsychosocial implications of scientific progress in the field of life and health sciences.

We could ask ourselves: why a new book on Bioethics when there are already numerous publications? We believe that a work that contemplates the discipline with a broad vision, anchored in its origins and with a vision of the future can contribute to a more global knowledge of a discipline that is beginning to go beyond the sphere of specialists to become an important element in modern societies.

The objective of *Bioethics: Foundations, Applications, and Future Challenges* is to present and debate the discipline in a trisynergistic way: from its foundations to the current debates concerning clinical and social bioethics, and from these to the future challenges that are glimpsed in the horizon.

This book is an updated approach to a discipline of special importance in our changing world. It presents the epistemological and historical bases that allow for establishing the disciplinary framework from the beginning. The chapters dedicated to the analysis of how bioethics is used to analyze complex situations in the biomedical field allow us to understand its importance to solve them and protect the human being. An important contribution of the book is the consideration of the current and future challenges that will require ethical consideration to ensure that biomedical research is not divorced from human rights.

The book is structured in three parts. In the first one, *The bioethical principles: state of art*, the appearance of bioethics, its epistemological approach, and its historical aspects are reviewed. It begins with the chapter *Bioethical and its epistemological status. A question of definition* where an analysis of the origin of the discipline is proposed, present in Greek philosophy and its differences with medical ethics, the first application of moral philosophy. The definitions of bioethics, its philosophical roots, and the relationships between bioethics and other disciplines are problematized, basing the concept of dignity.

The chapter *History of bioethics* reviews the origin of ethics in ancient civilizations, the birth of bioethics linked to medical care, as well as its emergence at the beginning of scientific medicine, and the background to the Nuremberg code. It also includes the bioethical developments linked to research and the use of human beings in unacceptable research procedures, clinical care, and the need to establish limits and rules to avoid medical paternalism, to later conclude with the recognition of bioethics as a separate discipline beyond medicine.

In the chapter *The principlist approach in bioethics*, the principlist approach in bioethics is exposed with special emphasis on the approach of the four principles formulated by Beauchamp and Childress (1979). The chapter describes the principles and their process of specifying and weighing when they are in conflict. Besides, the most relevant criticisms of principlism and its alternatives are analyzed.

The second part of the book, *Applied bioethics,* deals with the main applications of bioethics in the contemporary world, and how it has shaped the approach to biological, medical, and health policy challenges.

The first chapter deals with *Bioethics applied to biomedical research* and considers ethical issues in research based on the Nuremberg Code, the Belmont Report, and the Declaration of Helsinki. Its focus is on the concept and practical application of informed consent and the development of bioethical research committees, the regulation of clinical trials, and animal research. In addition to analyzing the historical journey of bioethics applied to research, the need to establish laws and protocols that safeguard human dignity and the conditions to participate in research as research subjects at all times is framed.

The second chapter on Applied Bioethics, *Unresolved issues in bioethics,* deals with one of the main unresolved issues in the discipline, such as the value of human life and the ethical legitimacy of its intervention, as well as it's beginning and end definitions. In particular, the chapter focuses on the ideological conflicts linked to the definition and value of human life. Far from being a global consensus, nor expect to be, the religious and legal debate on abortion and euthanasia, and other conflicts at the beginning and end of life are addressed.

The next chapter, *Ethics of Health Care Allocation of Resources. The Case of Organ Transplantation,* addresses health priorities, social inequalities, and political priorities when deciding on the allocation of economic and health resources. The theories of distributive justice are developed from moral philosophy to later analyze their role in practice. Particularly, the case of organ transplants is analyzed. The concept of brain death, its indicators, and the ethical problems are described, the criteria for the allocation of organs for transplants are analyzed, and the models of different countries to obtain informed consent for post-mortem donations are described. Likewise, the ethical aspects of living donor transplantation are analyzed, such as donor coercion and organ purchase.

The next chapter, *Healthcare priorities,* addresses health priorities, social inequalities, and political priorities when deciding on the allocation of economic and health resources. The theories of distributive justice are developed from moral philosophy to later analyze their role in practice. Particularly, the case of organ transplantation is analyzed, and the concept of brain death, its indicators, and ethical problems are described. The criteria for the allocation of organs for transplants are analyzed, and the models of different countries to obtain informed consent for post-mortem donations are described. Likewise, the ethical aspects of living donor transplantation are analyzed, such as donor coercion and organ purchase.

The chapter on *Social Bioethics* addresses the definition of this field, and the importance of the concepts of vulnerability, care, empowerment, social justice, and recognition. Although clinical bioethics occupied a preponderant place in the discipline for many years, current reflections on the need for professional approaches to include the social dimension make this approach essential. Likewise, the chapter reflects on global bioethics, the ethical treatment of animals, and environmental ethics, understanding that in all these areas there are beliefs and social attitudes involved.

In the third part of the book, *Future challenges*, some of what we believe will be the future challenges of Bioethics are addressed, although we will surely be surprised by some techno-scientific advance or unexpected social or other problems that we must analyze using an approximation. This part is devoted to the potential role of discipline in dealing with new situations.

It begins with the chapter *Challenges in biomedical areas*, dedicated to the challenges in biomedical areas. Technological developments and bioethical dilemmas related to reading and editing genetic information, CRISPR-Cas9 technology, biobanks, Big Data, proposals for human enhancement, and artificial intelligence are addressed. As mentioned before, techno-scientific advances do not stop and it is necessary to establish a dialogue between scientists, health professionals, and society as a whole, to establish the limits and possibilities that this innovation entails.

In the next chapter, *Roles and challenges for clinical ethics committees and clinical ethics consultation systems*, the new roles of bioethics committees are discussed with their expansion of functions beyond biomedical research, and how member training can be improved of the committee. Clinical ethics committees play an important consultative role in healthcare centers, and this chapter includes the historical background of two cases of committee interventions, and their functions and characteristics.

Inspired by the global crisis generated by the COVID-19 pandemic in the years 2020–2022, the next chapter, *Critical bioethical approach to health crisis scenarios*, analyze bioethical principles in health crisis scenarios. The purposes of health and human rights are analyzed, as well as the theoretical limitation of bioethical principles to protect the human being and the implications of rereading the principles from complex realities such as health crises. In addition, the chapter offers an analysis of bioethics in the Latin American reality, where there is a great opportunity to consider and try to solve new and old problems related to life and society, taking into account the weaknesses and strengths of social institutions and healthcare in that region.

In the last chapter, *The challenge of teaching bioethics*, teaching bioethics is addressed and a definition of the contents that should be taught is proposed. Taking into account the historical development of bioethics, and the growing challenges facing the discipline, the ideas that are sought to be transmitted and how bioethics is taught have also been changing over the years. The chapter analyses the pedagogical methods and didactic objectives of the main approaches and methodologies, and proposes some practical examples, concluding that teaching bioethics must be integrated into the curricula at all levels of schooling, even in fields far removed from health sciences, to develop responsible citizenship that be able to make rational decisions about the challenges facing our world.

This book is committed to the conception of bioethics as a transversal discipline in the life sciences. This is the background that led the Universitat de Vic – Universitat Central de Catalunya (UVic-UCC) to create in 2015, in collaboration with the Victor Grifols i Lucas Foundation, the Grifols Foundation *Chair in Bioethics*, coordinator of this publication.

The purpose of the Chair is to promote knowledge in the bioethics field through teaching, research, and transfer of knowledge. It is about promoting within the University and in a transversal way, a space for ethical and interdisciplinary reflection to introduce it in all lines of work and research such as health sciences, social services, biotechnologies, environmental sciences, and biomedicine, among others. The transversality of bioethics forces us to think, within the educational and professional field, about the ethics associated with scientific advances and their social interest, taking into account respect for life, the person, dignity, diversity, responsibility, and freedom.

The Grífols Foundation *Chair in Bioethics* was created with the mission of generating, disseminating, and transferring knowledge on the different areas of study of bioethics and delving into the ethical questions that are constantly raised by the life and health sciences. It disseminates and broadcasts the most current ethical issues, both nationally and internationally, to increase interest in this discipline and also aims to train a wide range of professionals who can face the new challenges posed by bioethics, offering spaces that favor collaboration between internationally recognized researchers and professionals with researchers and professionals from the UVic-UCC and the Grifols Foundation.

From our UVic-UCC *Chair in Bioethics*, we contribute to the challenge of ensuring that interest in ethical commitments goes beyond the strictly professional limit and is recognized by the societies of the 21st century. We hope that readers will join this essential challenge for the improvement of humanity and life on earth. The title of this introduction, borrowed from a well-known phrase by Sören Kierkegaard, insists on the need to know where we come from to progress in a changing future that we must face without any excuse.

Irene Cambra-Badii
Ester Busquets-Alibés
Núria Terribas-Sala
Josep-E. Baños

Contents

The Bioethical Principles
State of Art

Chapter 1
Towards a Definition of Bioethics as a Discipline

Carlo Orefice

Introduction. Why bioethics?

The field of bioethics has been difficult to situate in society and slow to take hold. Philosophy has at times rejected it, considering it "dangerously" close to contingent issues. Jurisprudence and politics have often deferred to the commonly held view that ethics should be confined to individuals' consciences. Indeed, in Italy, as in other countries, bioethics first appeared in a few universities in the 1990s. In recent decades, however, bioethics has slowly but progressively spread and grown in importance. These gains are only to a small extent attributable to the strengthening of bioethics as a discipline. Factors that have had a much greater impact on the importance of the field are related to the interest aroused in the social fabric in relation to a series of radical conflicts that have emerged from important discoveries in fields such as neuroscience and genetics and the real and potential applications of the knowledge acquired.

Scientific discoveries have led to new perspectives on the fundamental issues of life, and the accelerating rhythm of scientific discoveries and technological innovations in the twentieth century has meant that individual and collective consciences must adapt with equal urgency to this pressing rhythm.

Advances in science and their applications in life can be unsettling. As is evidenced by the succession of sensational news events, the growing expectations of well-being and health created by these advances is

Università di Siena, Italy.
Email: carlo.orefice@unisi.it

accompanied by crucial questions about the legitimacy of new interventions and procedures. Bioethics seeks to give society the tools to evaluate the goodness of medical and biological interventions as the border between what is technically possible and ethically acceptable becomes increasingly blurred (Giuliodori 2000).

Genetic research, pharmacogenetics and pharmacogenomics, research with human beings, and developments in biochemistry and molecular biology reflect medical and scientific progress, but they also raise numerous bioethical questions about human life. What is a person? In other words, how should we "define" a person? What characteristics distinguish a person from a non-person? What does it mean to respect the dignity of human beings? Ethical questions, amplified by the media, demand answers, especially on a clinical level.

On the other hand, bioethics has suffered from a fragmentation of viewpoints. Jurists, politicians, doctors, moral philosophers, and biologists deal with different aspects of bioethics in different capacities. A unifying cultural context and way of communication is lacking, resulting in specific reflections about the issues when approached from different fields. Technology is the cornerstone of modern medicine (Vilcahuamán and Rivas 2017), and progress in medicine depends on progress in technology. Medicine is neither a monolithic nor an exact science; rather, it encompasses an increasing number of specialties based on experimental sciences. Nevertheless, medicine is also the art of caregiving, and healthcare professionals are charged with serving vulnerable *Homo patients*. The fundamental mission of medicine remains the treatment of diseases and the alleviation of suffering, based on the principle of equity and justice.

This person-centered perspective reinforces the need for the humanization of medicine, animating and sustaining the idea of scientific progress itself and the statute of medicine (Spinsanti 1999). The definition of human life is already controversial, as is evident in arguments about when it starts and when it ends. Does personhood disappear when some essential capabilities are no longer present? Is it possible to limit personhood to individuals who possess certain abilities, on the basis of a rationalistic and functionalistic assumption? The question of personal identity is reflected in the doctor-patient-family relationship, as an emblematic expression of the human condition marked by vulnerability, and this is another central theme that interests bioethics.

Ultimately, bioethics is the place where radical conflicts are condensed.

After these initial reflections, we will attempt to define the field, outlining its epistemological status and discussing its still very controversial and unstable boundaries. Finally, we will examine the methods used in bioethics and attempt to identify the fundamental questions in the field.

The interdisciplinary nature of bioethics and lack of a single competent authority can be challenging, with input from multiple fields of research. However, it is in this *limes* that bioethics is its most innovative and productive. The discipline cannot be confined to a narrow disciplinary perspective; rather, its very essence involves posing concrete problems that can only be approached through knowledge from different disciplines.

In this sense, bioethics must not be relegated to worlds of healthcare and academics. Bioethics must be embraced by the entire whole community, because the issues involved are of the utmost importance for the fate of humanity (personhood, the meaning of life and death, procreation, family, solidarity, respect for the environment, to name but a few). These issues must be faced with continuity and competence, and each generation must learn about them in light of the transformations taking place. Society needs to acquire the tools necessary to evaluate bioethical issues and to consider the most appropriate choices of action.

The promotion of such a *bioethical culture* can help us to define the (fleeting) boundaries of bioethics.

The definition of bioethics in its epistemological aspects: is it a discipline?

What is bioethics and what is its epistemological status? Bioethics (from the Greek: ethics of life) is defined as "a new discipline that combines biological knowledge with a knowledge of human value systems in an open-ended biocybernetic system of self-assessment" (Potter 1975:2299).

Thus, to avoid vagueness in delimiting the epistemological status of bioethics, we need to reflect on the nature of medicine. The term "medicine" encompasses many meanings. On one hand, it refers to the enormous body of knowledge and set of techniques and healing practices used to preserve or restore human health, as well as to the research activity on which these are based. On the other hand, it refers to the use of these elements to treat human beings.

In the nineteenth century, Western medicine began to systematically apply the experimental method and thereby took on the fundamental characteristics of natural science. Thus, medicine is now considered an applied biological field of science that studies the composition, structure, and normal functions of the human organism, and it analyzes pathological phenomena affecting the organism to determine their etiopathogenesis and structural alterations with the aim of preventing, diagnosing, and treating disease. Physicians' bedside activity, often referred to as medicine, might be more accurately named clinical medicine or simply clinical activity. From an epistemological point of view, it is this clinical activity that raises the most interesting problems for bioethics.

In fact, what is the authentic nature of clinical activity? Is it a scientific activity similar to all other natural sciences? Is it a technology that applies notions drawn from the general sciences? Is it a purely professional activity? Or is it a special research activity in which the analytical study of individual cases leads to the formulation of laws and general theories? Each of these hypotheses has found supporters and opponents, but views of the nature of medicine have always been partial and one-sided.

Epistemologists have divided the various sciences into three basic categories (Antiseri 1975, Cosmacini 2011, Porter 2011): (a) the pure or theoretical sciences, which aim to identify regularities that constitute the laws of nature or to interpret different phenomena by collecting them within those complex systems of relations that make up scientific theories; (b) the technological sciences, which aim to predict phenomena to find ways to solve relevant practical problems; and (c) the historical or idiographic sciences, which aim to explain individual facts or phenomena at specific points in time and space.

Medicine comprises many different sciences, each of which falls into one or more of the three categories described. For example, physiology and pathology are two pure sciences that aim to describe and explain the phenomena that occur in normal and sick organisms, respectively. Pharmacology is a technological science that aims to identify substances that favorably modify the course of pathophysiological phenomena and clarify their mechanisms of action. Although clinical medicine can in many instances be considered a technological science, it is nevertheless primarily an idiographic science that focuses on diagnosing and treating patients rather than on elaborating new theories on pathological phenomena or pharmacodynamics. Clinical medicine seeks to explain the causes of a particular patient's ailments to predict and modify their future course. To achieve these goals, clinicians use the laws and theories of the pure and technological sciences. They describe their patients' pathological phenomena and explain them in light of knowledge from the general biomedical sciences. Anatomy, physiology, biochemistry, microbiology, genetics, and pharmacology provide clinicians with the concepts that enable them to understand their patients' disease.

Since medicine comprises many scientific disciplines of different natures, it cannot be considered a pure science, exclusively devoted to knowledge (such as cosmology or zoological taxonomy). Rather, medicine is a science that has a very specific purpose. If medicine were interested only in knowledge, it would not devote so much attention and research to issues that do not appear particularly important from a theoretical point of view. Disciplines such as semiotics, surgery, clinical pharmacology, and hygiene are central to medical science, but their importance is fundamental only in relation to the aims that doctors strive to achieve.

In short, medicine is an applied science, in other words, a science that is intimately connected to a specific purpose it intends to achieve (Antiseri 1975).

Thus, medicine is primarily an applied biological science, but is it *only* an applied biological science? Until the mid-twentieth century, the idea that medicine was nothing more than a scientific discipline was the dominant view, although many developments had started to raise questions about such a simplifying vision. One of the most important of these was the change in the patient's role in the doctor-patient relationship (Good 2006). While in the first half of the twentieth century the doctor established both the goal of the medical act and the means to achieve it, in the second half of the century patients became increasingly aware of their right to establish the ultimate purpose of the clinical act, as well as to choose the means that they deemed best to achieve that end. This change in the patient's role has also had a profound effect on the concept of the medical act: while earlier it was believed that the discourse on means and ends constituted a unicum, which was part of the scientific discourse, more recently doctors have become aware that the ends and values involved in their actions do not fall within the scope of scientific reason.

Another fundamental element that added complexity to the image of medicine in the second half of the twentieth century is its relation to technologies. Indeed, this element is so important that it affected the very foundations of the concept of medicine. And it is in this context that the bioethical discourse in the biomedical sciences gradually takes shape. Bioethics arose when scientific progress had become overwhelming and deontologies no longer seemed sufficient to deal with the issues that arose; in other words, it became necessary to reassess major scientific orientations in the face of new and demanding issues that involved complex and unprecedented situations and ethical problems. Organ transplantation, gene therapy, cloning, the use of stem cells, pre-implantation diagnosis, transgenic technologies, and other developments in the past half-century led to a real "revolution" in biomedical ethics. The transformation that took place was sudden and radical, changing conceptions of the relationships between those who professionally provide care and those who receive it, as well as the way that ethical rules concerning information are managed. These changes affected everyone: health professionals, patients, and patients' families.

In Western countries, these transformations have been the cause and consequence of substantial changes in social evolution, with significant repercussions on ethical feelings and personal behavior. On the one hand, the rapid secularization that took place all over the world, with the progressive emancipation of culture, customs, and social life from the once decisive influence of religion, has fueled a process of differentiation and multiplication of religious, philosophical, political, and cultural orientations. On the other hand, in the late twentieth century, the crisis of ideologies (i.e., the loss of influence of the utopian or conservative, revolutionary or reactionary worldviews that had characterized the first post-war period as well as the protests of 1968) led to the exaltation of both personal and class rights (e.g., women's movement), as well as of ecological animal rights movements, among others.

Medicine's incorporating new technologies brought progress in every care sector.[1] In fact, at the end of the 1980s, it was clear that scientific-professional and technological skills were capable of influencing moments then considered "natural", and as such not manipulable (e.g., the beginning and the end of human life), moments that have always been imbued with very high anthropological, psychological, religious, philosophical, and juridical meanings to the point of characterizing human cultural development. This exponential progression of technological applications has progressively created an exceptional and unprecedented concentration of hopes, expectations, and anxieties about questions related to the human body and, more widely, to the psychic, spiritual, and physical unity of the human being.

Against this background, with a sense of urgency and in the hope of articulating reflections that might aid the rational analysis of emerging moral problems, a new field that would come to be known as "bioethics" started to emerge in 1970/1971 in the USA and in some European countries. These initial developments were followed by the birth of *another* bioethics that is closer to the experience of ordinary people. This "bioethics of everyday life" (Berlinguer 2003) takes into account bioethics' relationships with democracy and politics, underlining the social dimension of health; more than biology alone, health must also be linked to social struggles and labor movements and their relations with the ruling classes (Foucault 2021). This principle of self-determination in issues concerning health and life (e.g., euthanasia, experimentation with drugs and therapies, etc.) has brought about a growing awareness of the impact that choices related to bioethics can have on daily life.

The bioethical questions of the previous half century have fallen between two extremes: the first, in which a variety of solutions could be used for long-standing, fundamentally traditional moral problems, and another, in which exceptional and previously unthinkable advances in

[1] Anti-infective drugs (e.g., antibiotics), anesthetics and analgesics, immunology (organ transplants), birth-control pills, artificial insemination, and *in vitro* fertilization. A few years after Watson and Crick (1953) described the structure of DNA, genetic biotechnology advanced to the first gene transfer activities (Anderson 1984) and the start of the recombinant DNA industries, enabling the production of vaccines and therapeutic proteins, etc. as well as the application of genetics in humans, especially in the diagnostic sector.

the tools made available by science and technology have given rise to unprecedented complexity in the questions themselves as well as in their solutions.

All these aspects are points of emerging fracture in the West, which are ultimately attributable to divergences that are often related with ontological, metaphysical, and anthropological criteria as well as with moral criteria.

Bioethics and interdisciplinarity

From the creation of the discipline of bioethics, its boundaries have been blurred.[2] Bioethics is a territory that belongs "to everyone and to no one". It arose from applied ethics and is naturally inclined toward moral problems, but it also grew out of oncological medicine and has close ties with biology, biomedicine, bioengineering, and biotechnology. Thus, it was not particularly configured as a discipline, but rather as a point of convergence between multiple fields of research, *limes* between ethics, medicine, and life sciences. This interdisciplinary nature makes it a "no man's land", and it does not seem possible to identify a single competent authority. However, at the same time, this interdisciplinarity is one of bioethics' greatest strengths, preventing it from being locked in a narrow disciplinary perspective and allowing for innovation. By posing concrete problems that involve knowledge from different disciplines (e.g., biology and natural sciences, sociology, ethics, law, religion, among others), bioethics requires multiple inputs from different perspectives.

As Potter (1971) correctly intuited, bioethics must serve as a bridge between different types of scientific and humanistic knowledge. Exploring the issues addressed by the discipline requires a constant and challenging interdisciplinary dialogue between representatives of different disciplines, schools, and perspectives, in the search for a possible shared outcome. We import into bioethics, and we adopt (not without arguments and mediations) the fundamental philosophical references that orient our

---

[2] The official date of birth of bioethics can be traced back to 1971, the year of publication of a book entitled *Bioethics. Bridge to the future* by V.R. Potter, a medical oncologist at the University of Wisconsin School of Medicine, who highlighted the causal relationship between the environment and cancer. In this book (a collection of 13 articles, 10 of which had already appeared in various American scientific and economic journals between 1962 and 1970), Potter denounced the division between the scientific and humanistic fields of knowledge as unnatural and dangerous and advocated a "bridge" between these two cultures. Following this rigid separation of roles and competences, humanistic culture had claimed ethics as its own, exclusive terrain, thus depriving science of a guide for action that includes moral values. But, according to Potter, ethical values should not be separated from biological facts: bioethics, as a new science, should therefore "bridge" these two cultures (Chiarelli and Gadler 1989).

intellectual schemes. And here the difficulties begin, because the different schools hold to different fundamental philosophical viewpoints.[3]

Since the "new" issues that bioethics deals with come into being through the development of biotechnologies, it is essential to find a middle road between absolute trust in science and severe aversion to science (or between technophilia and technophobia) and to strive to achieve a balanced view. From this position, the fundamental challenge concerns not only, and perhaps not even in the first place, the solution of individual problems with their dense moral dilemmas, but rather seeking the fundamental balance that must necessarily exist between the advancement of scientific knowledge and related technological applications on the one hand, and a deeper meditation on life, humanity, and the ecosystem, on the other. It is at this crossroads that bioethics intersects with other disciplines.

These considerations raise the question of whether bioethics can exist alone or whether it is completely bound to other fields of knowledge. To answer this question, it should be emphasized that bioethics adopts a gaze that is not identifiable with that of other sciences, in the sense that it does not acquire the certainty of existing from science, but rather the need to constantly reflect on its own field of interest. Despite the variety of methods, principles, and approaches that tempt one to speak of "bioethics in the plural", bioethics is fundamentally a philosophical reflection.

The Latin American author José Kuthy Porter considers bioethics a derivation of philosophy *tout court* (not only from ethics), when he maintains that "for my part, I conceive bioethics as a new place of encounters between science and humanism (understood as a revaluation of the human being), between science and clear conscience: and it is for this reason that its concrete application must be in function of the spiritual enrichment of the person in the environment that surrounds him. [...] by deriving bioethics from philosophy it is possible to understand how in Latin American countries the development of bioethics is encountering a methodological drawback that must be pointed out, and which consists in the fact that philosophy itself has not developed sufficiently in the Continent" (CNB 1995:72).

Warren Thomas Reich, on the other hand, looks towards ethics, rather than the entire sphere of philosophy, when he observes that "the word bioethics, commonly used as a slightly more extensive medical ethics, that is, as the equivalent of the biomedical ethics of the relationship between

[3] For example, the schools of ontological personalism, utilitarianism, the relational approach, the approach according to values and/or rights, contractualism, empiricism, phenomenology, and numerous others that attract attention and discussion; all these approaches are operating in many bioethical arguments.

doctor and patient and between researcher and subject, and therefore like the ethics of the health professions, soon took on a much broader meaning and precisely that of the ethics of the sciences of life. It [the word 'bioethics'] conjugated the environmental sense that van Rensselaer Potter had in mind when he first coined the term, with the health connotated inspiration of the Kennedy Institute of Ethics, when it was first conceived, with the aim of using the word in an institutional setting. So, by bioethics, we mean the ethics of health, health care, and all sciences concerning life. In my view, bioethics includes not only the ethics of medicine and other health professions but also the ethics of biomedical research, public health, environmental hygiene, human reproduction, demographic policies, the care of animals and of the whole environment" (CNB 1995:77).

Whichever position we adopt, the term "bioethics", as explained above, appears disorienting if too restrictive, since the reality and the reach of the questions that it deals with must be shared with other disciplines outside philosophy. Indeed, bioethics involves more than a diversity of perspectives or ranking of values in bioethical issues—bioethics comprises humankind itself in its multiple physical, biological, and anthropological meanings.

Compared to the problems briefly referred to, the anthropological question has more radical characteristics and appears destined to become more and more pervasive for two reasons. First, because new technologies affect the subject and transform it, they tend to change the way we understand the central notions of our individual experience (being generated or produced, being born, living, procreating, seeking health, growing old, dying, etc.). Second, what is called into question is not so much an idea of a body bound by the laws of anatomy and physiology, but a construct that is at the center of practices, discourses, imaginations, and representations that are social and cultural (Orefice 2016). In other words, individuals exist not a priori, on the basis of some abstraction, but because they are inserted in a social context. Individuals' interactions with their environment (physical, cultural, and social) allow them to acquire the skills necessary for creation and production of intelligible social episodes (Harré 1994). Therefore, rather than defined in exclusively natural terms, our identity is a social product whose construction processes are to be investigated: whether we are talking about the ways of self-representing health or disease, the ideas we have about what life or death is, of transgenic technologies, the body appears as an extraordinary vector of understanding—or lack thereof—the relationship between the individual and the world.

Starting from this recognition of the socio-historical nature of the body, while bearing in mind that it is also active in producing cultural meanings and experiences (Remotti 2002), we can abandon the idea of a body given "in nature" and begin to use interpretations of contemporary medical science that look critically at the medical system by incorporating underlying anthropological concepts (Spinsanti 1991, Masiá Clavel 2004, Pizza 2015, Tolone 2016).

The various bioethical questions evoked here require a rethinking of the anthropological question and lead to the formulation of a further question that appears essential for understanding not only the relationship between bioethics and other disciplines, but also the epistemological status of these disciplines: why does modern medicine fail to consider the human being as a whole?

Indeed, conceiving the body as a non-culturally constructed reality results in a classificatory, analytical, and reductionist view (Sahlins 2010). Numerous authors (Farmer 2003, Quaranta 2006, Zannini 2008) have pointed out that biomedical science provides but one, specific representation of the body, among many. This representation is certainly effective and legitimate, but its strength lies mainly in the degree of universalism that it is capable of containing, rather than in the reductions it operates by often arrogating the right to judge other cultural systems absolutely. The anthropologist and physician Michael Taussig (1980), in an old but always current article, highlights how clinical reasoning works on the basis of a process of reification through which human relationships, people, and experiences are objectified as mere "facts of nature". For Taussig, biomedicine becomes a means of social control, functional in the conservation of a particular structure politics by eliminating the social, economic, and political dimensions incorporated in the disease, and disguising itself "as a science of (apparently) real things" (1980:3). In this perspective, the scholar's task appears to be to reveal what lies behind medical ideology, demystifying its constructions. As each individual may experience, the disease does not reside only in an alteration in the structure and/or functioning of the individual biopsychic organism (disease), but also in the reasons for the pain (Le Breton 2007), an experience difficult to forget, making it personal (*illness*), as well as social and political (*sickness*). It suffices to go back to the knowledge of the body found in popular traditions, or in non-Western societies, to see how much knowledge of the body exist and how the body is always "open to the world" and that it is not cut off from the subject or isolated from the cosmos (Diamond 2012). It suffices to recall the various grotesque mystifications of medical practice, in which we all struggle to witness how our body is not just an organic mosaic of biological entities, but a "construct" crossed by social and political forces that produce specific discursive devices that found regimes of historically determined truths (Orefice 2013).

In trying to explain why modern medicine fails to consider the human being as a whole, we find ourselves entangled in (medical) knowledge that conceives man *in the abstract*, where the organs that make up the body are—conceptually and methodologically—separated from each other. This "myth of simplicity" (which controls the adventures conceived by Western thought from the seventeenth century onward) has been extraordinarily fruitful for scientific knowledge: by isolating the patient's body from its cultural context, this knowledge has focused medicine's gaze, enabling it to concentrate on controllable and predictable factors. Nevertheless, the resulting "anthropological hole" is completely evident.

Therefore, it can be useful to look at bioethics and its relationships to other disciplines when this examination results in the proposal of a "system theory" centered on the combined physical, biological, and anthropological aspects of the body that does not further reduce or attempt to simplify the phenomena of complex organization, instead considering the units of physics, biology, anthropology "open". Thus, bioethics requires an interdisciplinary perspective that "reintegrates" those uncertain, ambiguous, and contradictory aspects of the different disciplines—each with its own specificity—in considering the central problems of human experience and action.

Basic methodological elements of bioethics: Some considerations

According to the first edition of the *Encyclopedia of Bioethics* (1978), the fields of investigation of bioethics concerned the ethical problems of all health professions, behavioral research regardless of its therapeutic applications, health policies, occupational medicine, population control policies, and the problems of animal and plant life in relation to human life (Allahan 1995). In the second edition, more than 15 years later (1995), this delimitation no longer seemed sufficient, so much so that in the chapter entitled Bioethics Daniel Callahan (co-founder of the world's first bioethics research institute) highlighted the difficulty of defining a constantly evolving field. The evolution of the field of relevance of bioethics, in its multiple angles, has been driven by its focus on problems of importance for the fate of humanity. Problems of this type require continuity and competence. New generations must be made aware of the transformations taking place, and the current generation must help them develop the evaluative tools to make any choices.

In structuring itself as a "bridge" between the scientific and humanistic fields of knowledge (Potter 1971), bioethics has progressively evolved from being the exclusive prerogative of the healthcare world to penetrate law and politics, philosophy, literature, pedagogy, social sciences, and the mass media. This expansion has favored a *bioethical culture*, opening the possibility of education in bioethics beyond the confines of the academic world to include the community as a whole. Themes such as the person, the meaning of life and death, procreation, family, solidarity, respect for the environment, to name but a few, have always been an integral part of the individual's wealth of information and training: the novelty lies in discussing the interpretation of these phenomena due to the evolution of scientific knowledge and the use people derive from it.[4]

In both bioethics limited to the health/academic world and bioethics extending into everyday life (Berlinguer 2003), it seems necessary to consider the relationship between education and bioethics and specific training in the discipline. The risk is that without adequate pedagogical reflection, that is, without an elaboration of ideas and actions that provide a view that works toward the integral promotion of the person according to the anthropological vision, education becomes devoid of meaning and diachronicity (ten Have 2015).

Fostering a bioethical culture beyond the academic world entails more than simply informing individuals about health problems to direct them to primary and secondary prevention, modifying habits and lifestyles, or presenting the achievements of technical science. It is important to make people aware of the problems and encourage them to participate responsibly in bringing about their own wellbeing and the common good. To achieve this goal, education must distance itself from "pre-understandings", feelings on which life choices are based, training events, ethical judgments, and the need to take a stand with regard to everything that science conquers and proposes, to reach an end that guides and motivates certain choices.

In the world of healthcare, university training in bioethics meets two main needs and intercepts a problem. On the one hand, there is a need to reflect on the ethics of training, a trait that should distinguish every educational approach and should underline the integral and not optional role of ethics in the healthcare professions; ethics must find an adequate translation in training and consequently permeate ordinary care practice (Loro 2008). On the other hand, there is a need to pay specific attention to bioethics training in individual scientific disciplines: in other words, certain formative actions and specific preparation are needed during specific time periods of human life, from birth all the way to death, inevitably touching the meaning and significance of medicine today. The problem relates both described needs: what methodological choices can adequately support bioethics, understood as an analysis of emerging moral problems? And this question leads to another: is an approach based on theory better than one based on the analysis of specific cases?

[4] It should be emphasized that bioethics, in addressing problems that transcend interpersonal relationships, also leads to reflection on the responsibilities we have toward living organisms and the processes of nonhuman life (animals and plants included), promoting an "ecological mentality" that invests and affects a plurality of scientific fields.

Bioethics[5] is not indifferent toward the answer to this question because the teaching methodology does not represent an "accessory element, which can be easily renounced, but an essential channel to stimulate a greater and better understanding of the problems and to facilitate an internalization process that solicits in the students a rigorous reflective attitude, which helps them in questioning themselves personally, analyzing their attitudes, their convictions and their dispositions" (Nordio 1997:140). If, therefore, it would seem that "bioethics must be reduced to a discipline in order to be taught, we should not lose the deep conviction that it is a cultural movement that brings knowledge and ideas while problematizing and dragging feelings with itself" (ibid.). For professionals, a method based only on case analysis has a number of limitations: in fact, relying too much on case studies can increase controversies in the moral field; it is often not easy to identify the specific problem that has moral relevance; some questions that need to be confronted go unasked and others should not be asked.

This is one of the most complex aspects of bioethical discourse, and it would require greater study even before single issues of bioethical interest were to be addressed. It would seem insufficient to solve the supplementary need after bioethics experts have acquired heterogeneous skills, starting from their own discipline; moreover, merely coordinating with other disciplines to enhance the contributions of each could not overcome this insufficiency. Indeed, this approach could entail an additional risk: the lack of real integration between fields of knowledge could result in greater difficulty for students whose perspectives are as yet unformed. To overcome the shortcomings of the traditional training system, which could lead to early conditioning toward a scientist's mentality, health workers might be offered many analytical tools that they are unprepared

[5] This is a difficulty shared with the medical humanities themselves, of which bioethics is an integral part. As for the medical humanities, if the aim is to prepare health workers to face and solve complex professional problems that they will most often encounter, not only from a technical-scientific point of view, but also from a more broad and articulated view that takes into account all the components of these problems (from psycho-relational to ethical components, from anthropological to philosophical and epistemological components, from socio-cultural to historical and economic components), these objectives are "hardly formalizable in notions and theoretical knowledge, even if they are nourished by these as well [...]. Therefore the objectives of the medical humanities cannot be only those of knowing abstractly the characteristics of the functioning of the psyche or the techniques of effective communication, the theoretical rules of logic, the abstract principles of ethics, or the sequence of fundamental events in the history of medicine, or the laws of health economics, or the history of literature, cinema, philosophy and art" (Vettore 2005:1018), but they strive to overcome the conflict between hard culture (the natural sciences) and soft culture (the human sciences), orienting toward an integration of knowledge and bringing the patient and health worker back to the center (Orefice and Baños 2018).

to synthesize. By replacing the concept of multidisciplinarity with the concept of interdisciplinarity, we make it clear that the static nature of a juxtaposition must be replaced with the dynamism of convergence of different disciplines that, even if with different epistemological statuses and different research approaches, must be integrated into a unitary epistemological and value-based perspective.

Thus, theoretical analysis in a process seeking a *reflective equilibrium* (Brody 1990) combined with coherence through case studies and theory seems to be a solid methodology for approaching the need to deepen the founding aspect of bioethics. Otherwise, there is a risk that bioethical reflection might hypertrophy, growing horizontally without growing vertically.

Conclusions

Bioethics is rooted in the ontological dignity of humankind. Bioethical issues require us to delve deeply into the anthropological question. Professionals must move past self-sufficiency and acquire skills to favor a truly interdisciplinary space. Bioethics must open up to embrace a project of integral promotion of the person, where everyone is called to participate. The appeal to ethics must not be relegated to urgent, exceptional situations that are strongly charged with emotion. To achieve these goals, bioethics as an academic discipline and an interdisciplinary field of reflection must take advantage of the "privilege of proposing rather than imposing; it may indicate the goals to be achieved, rather than the chasms to be avoided; favoring synergies, rather than opposing ideology to ideology. And above all, it must know how to articulate and illustrate its reasons" (Spinsanti 1999:6).

In the health sector, the doctor-patient relationship supports the structure of scientific medicine and is a particularly significant starting point for bioethics. The implicit presupposition of every health action still appears to be the doctor explaining the disease, while the patient is unaware of what is happening inside of him (Good 2006). This approach is directly linked to the biomedical approach, which tends to "understand" disease only at the biological level, thus neglecting (or at least reducing) the complexity of the disease and becoming a succession of cause-and-effect events that neglect contextual and "systemic" aspects. Focusing treatment only on biological mechanisms can impoverish the "human dimension" and the doctor-patient relationship (Zannini 2008, Orefice 2020), relegating the "sick" to a situation of subordination and denying them a "right to meaning" that would allow their experiences of illness to be "explained" through a plurality of factors (cultural, economic, symbolic, etc.). The advent of highly sophisticated technologies has increased the gap between *explainable* symptoms and symptoms that cannot find an exclusively medical explanation, thus changing the doctor-patient relationship, which is increasingly mediated by instrumentation.

It seems clear, therefore, that one of the fundamental tasks of bioethics is to rethink the practice of medicine itself in the context of the needs of the entire human being, not mutilated or diminished. Only under this condition does it seem legitimate to think of a *different* medicine, and of bioethics that in expressing an anthropological question requires disciplines to shed their rigidity to face complex and fundamental problems. This approach to bioethics can become a regulating principle in education through which a wide variety of problems in health care can be dealt with and help favors a bioethical culture in the community as a whole.

References cited

Allahan, D.C. 1995. Bioethics. pp. 278–287. *In*: Reich, W.T. (ed.). Encyclopedia of Bioethics - Vol. I (revised edition). MacMillan Simon & Schuster, New York, USA.

Antiseri, D. 1975. Epistemologia e didattica della storia. Armando, Rome.

Berlinguer, G. 2003. Everyday Bioethics: Reflections on Bioethical Choices in Daily Life. Baywood Publishing Co., Amityville.

Brody, B.A. 1990. Quality of scholarship in bioethics. The Journal of Medicine and Philosophy 15: 161–179.

Chiarelli, B. and E. Gadler. 1989. Nota storica. Van Rensselaer Potter e la nascita della Bioetica. Global Bioethics 2: 61–63.

CNB - Comitato Nazionale di Bioetica. 1995. Bioetiche a confronto. Atti del Seminario di studio, 20 ottobre 1995, Presidenza del Consiglio dei ministri, Dipartimento per l'informazione e l'editoria, Rome, Italy.

Cosmacini, G. 2011. L'arte lunga. Storia della medicina dall'antichità a oggi. Laterza, Rome.

D'Agostino, F. 2003. Parole di Bioetica. Giappichelli, Turin.

Diamond, J. 2012. The World Until Yesterday: What Can We Learn from Traditional Societies? Viking Press, New York.

Farmer, P. 2003. Pathologies of Power. Health, Human Rights, and the New War on the Poor. University of California Press, Berkeley.

Foucault, M. 2021. Medicina e biopolitica: la salute pubblica e il controllo sociale. Donzelli Editore, Rome.

Giuliodori, C.G. 2000. Bioetica e comunicazione. pp. 117–128. *In*: Sgreccia, E. and M.L. Di Pietro (eds.). Bioetica e formazione. Vita e Pensiero, Milan, Italy.

Good, B.J. 2006. Narrare la malattia. Lo sguardo antropologico sul rapporto medico-paziente. Einaudi, Turin.

Harré, R. 1994. L'uomo sociale. Raffaello Cortina Editore, Milan.

Le Breton, D. 2007. Antropologia del dolore. Meltemi, Rome.

Loro, D. 2008. Formazione ed etica delle professioni: il formatore e la sua esperienza morale. FrancoAngeli, Milan.

Masiá Clavel, J. 2004. Bioética y antropología. Universidad Pontificia Comillas, Madrid.

Nordio, S. 1997. Riflessioni metodologiche: come si inserisce la Bioetica nelle Facoltà di Medicina? pp. 140–158. *In*: Cattorini, P. and V. Ghetti (ed.). La Bioetica nelle facoltà di medicina. FrancoAngeli, Milan, Italy.

Orefice, C. 2013. Per una pedagogia "di confine". Decifrare differenze, costruire professionalità. Edizioni Unicopli, Milan.

Orefice, C. 2016. Unicità e pluralità dell'essere corporeo. Educare alle differenze. pp. 59–70. *In*: Cunti, A. (ed.). Sfide dei corpi. Identità Corporeità Educazione. FrancoAngeli, Milan, Italy.
Orefice, C. and J.-E. Baños. 2018. Introduction. The humanities and medicine: why do they need each other? pp. 10–15. *In*: Orefice, C. and J.-E. Baños (eds.). The Role of Humanities in the Teaching of Medical Students. Dr. Antoni Esteve Foundation, Barcelona, Spain.
Orefice, C. 2020. Lo studio della cura educativa in un'ottica complessa. PensaMeltimedia, Lecce.
Pizza, G. 2015. Antropologia medica. Saperi, pratiche e politiche del corpo. Carocci, Rome.
Potter, V.R. 1971. Bioethics: Bridge to the Future. Prentice-Hall, New Jersey.
Potter, VR. 1975. Humility with responsibility: A bioethic for oncologists. Cancer Res. 35: 2297–2306.
Porter, R. 2011. Breve ma veridica storia della medicina occidentale. Carocci, Rome.
Quaranta, I. 2006. Antropologia medica. I testi fondamentali. Raffaello Cortina, Milan.
Remotti, F. 2002. Forme di umanità. Mondadori, Milan.
Sahlins, M. 2010. Un grosso sbaglio. L'idea occidentale di natura umana. elèuthera, Milan.
Spinsanti, S. 1991. Bioetica e antropologia medica. Carocci, Rome.
Spinsanti, S. 1993. Prefazione all'edizione italiana, pp. 5–12. *In*: Gracia, D. (ed.). Fondamenti di Bioetica. Sviluppo storico e metodo. Ed. San Paolo, Cinisello Balsamo, Italy.
Spinsanti, S. 1999. Le ragioni della Bioetica. Edizioni Cidas, Rome.
Taussig, M.T. 1980. Reification and the consciousness of the patient. Social Science and Medicine 14b: 3–13.
ten Have, H. 2015. Bioethics Education in a Global Perspective: Challenges in Global Bioethics. Springer, Dordrecht.
Tolone, O. 2016. Alle origini dell'antropologia medica: il pensiero di Viktor von Weizsäcker. Carocci, Rome.
Vettore, L. 2005. Si possono apprendere e insegnare le Medical Humanities? Med. Chir. 27: 1016–1021.
Vilcahuamán, L. and E. Rivas. 2017. Healthcare Technology Management Systems: Towards a New Organizational Model for Health Services. Elsevier Science, Amsterdam.
Weizsäcker, V.v. 1996. Filosofia della medicina. Guerini e Associati, Milan.
Zannini, L. 2008. Medical humanities e medicina narrativa. Nuove prospettive nella formazione dei professionisti della cura. Raffaello Cortina, Milan.

Chapter 2
History of Bioethics

*Josep E. Baños** and *Elena Guardiola*

Introduction

There is no general agreement to establish when bioethics appeared as an independent discipline. This fact is a consequence of the early connection of bioethics with medical ethics and deontology. The latter were linked with the birth of modern medicine in the Ancient Greece and physicians became more or less interested in them in the following centuries. An important element of these early times was the Hippocratic Oath which is still taught in many medical schools and is even a ritual part of their graduation ceremonies around the world.

Interest on bioethical issues recovered with the appearance of *Medical Ethics,* a book published by Sir Thomas Percival in 1803. In fact, this was the first time that the name of the discipline was used in the title of a book. It influenced the consideration of ethical aspects of medical practice in some societies, as explained below, but this interest was restrained only to the professional field.

Social awareness of the importance of ethical components in the practice of medicine should wait until the next century with the discovery of how prisoners from German and Japanese concentration camps were included in alleged biomedical research. Some of the people who performed the Nazi experiments (1942–1945) were tried in the Nuremberg trial, sentenced and, in some cases, executed. What was seen and heard those days had an important impact on Western society, which considered that these events could not be repeated. The most important regulatory consequence was the conception of the Nuremberg Code (1947), a

Universitat de Vic – Universitat Central de Catalunya, Spain.
* Corresponding author: josepeladi.banos@uvic.cat

document that established for the first time a worldwide agreement on how medical research should be carried out in order to respect ethical principles and protect participants.

Certainly, the Nuremberg Code contributed for the first time to the social recognition of the importance of moral aspects in the doctor-patient relationship, but only in medical research. Unexpectedly, some groups did not accept the validity of the Code as they considered that it was only a consequence of the unacceptable activities of Nazi physicians during World War II. In their opinion, these behaviours were not present in countries other than Germany. It was concluded that a new document was needed, and finally the World Medical Association prepared the Declaration of Helsinki (1964) that lacked these beliefs. This Declaration followed periodic updates in the following years but it could not prevent unethical actions in medical research. For instance, in the United States, several published studies showed that the principles stated in the Declaration of Helsinki were not followed and prompted to new actions that conveyed to the Belmont Report (1978).

A new situation arose in the 1950s with the appearance of medical and technological discoveries that put the ethical commitments beyond research with human beings to enter clinical practice. Some examples of these new dilemmas were how to establish the priority of patients in the use of haemodialysis in renal failure, prenatal diagnosis, abortion, euthanasia, *in vitro* reproduction, embryo selection or the use of DNA edition genomic techniques like Clustered Regularly Interspaced Short Palindromic Repeats (CRISPR). These new ethical challenges conveyed a new situation; a moral approach was recognized as an unavoidable need to consider the application of these technologies to human life. This moral consideration considered that ethical analysis was not restricted to medical professionals but also to other members of modern societies, such us philosophers, lawyers or theologians, among others.

This chapter reviews the historical evolution of bioethics with a detailed approach to those circumstances than helped shape the creation and development of a new discipline independent from medical ethics.

The origins of the name of the discipline

For many years, it was generally accepted that the US biochemist and oncologist Van Rensselaer Potter (1911–2001) coined the term bioethics in 1970 (Potter 1970). The main reason for this belief was the publication of his book entitled *Bioethics: Bridge to the future* (Potter 1971). However, Fritz Jahr (1895–1953), a German theologian and teacher, created the term 'Bio-Ethik' and used it from the mid-1920s (Jahr 1926, 1927). Jahr used it to refer to the acceptance of ethical duties not only towards human beings but also towards every living being, including animals and plants.

This was for him an extension from the Kantian categorical imperative to a bioethical imperative in order to protect a threatened planet. These contributions were ignored for at least five decades. In 1997, Rolf Löther (1997) mentioned the name of Fritz Jahr as the creator of the word Bio-Ethik in 1927; however, the analysis of Jahr's conceptualization arrived with Hans-Martin Sass one decade later (Lolas Stepke 2008, Sass 2007, 2008).

Jahr's conception was clearly different from Potter's proposal given the different historical setting. In fact, Potter did not take Jahr's previous work into account. The definition used by Potter considered the birth of a new discipline under the influence of the Cold War, the advances of medical research, the effects of the Nuremberg Code and the Declaration of Helsinki, the environmental threats and the conscious feeling that ethical considerations in medicine should not be restricted to medical professionals.

Muzur and Rincic (2015) analysed the conceptual and philosophical differences of Jahr's and Potter's definitions. In this way, Jahr's concept is closed to a global bioethics that considers the relationships of human beings with the planet, but does not include biomedical advances and their potential consequences for patients (Giovanni 2001). This is not unexpected, as medical advances were still at the beginning in the 1920s. A detailed analysis of Jahr and Potter's works can be found in Pessini's review (2013).

Fritz Jahr's proposal can be understood beyond the introduction of the word bioethics: his intention is to reflect on human and natural life, and the responsibility we have with ourselves and other species. Jahr reflects through cultural sources of the time, such as literature and opera, evidencing the humanistic richness of bioethics since its inception (Lima 2009).

This different conceptualization of the definition of bioethics followed Potter. Later, Dutch obstetrician André Helleger and political activist Sargent Shriver used this term again when the Joseph and Rose Kennedy Institute for the Study of Human Reproduction and Bioethics was founded at Georgetown University in 1970. For some authors, its meaning is closest to the current meaning of bioethics (Chadwick and Wilson 2018). Helleger and Shriver defined the discipline as the analysis of the ethical issues currently raised by medicine and the biological sciences. This meaning is far from Jahr's global view and Potter's environmental compromises; it focused the interest of bioethics in the advances of biomedical research and their consequences for medical practice in the second half of the 20th century.

Ethics in the old civilizations

Some authors consider that the Hammurabi Code, written around 1750 BC, is the oldest document that states some aspects linked to medical ethics. However, this code only comments peripheral aspects of the practice of medicine in some articles, like how to reward some medical activities or how to punish mistakes carried out by physicians. Therefore, it does not consider the ethical aspects of the medical profession or the moral consequences of its choices.

In the Western medicine, the first document that approached the intrinsic moral aspects of physician's activities was carried out by Hippocrates, the most well-known Greek physician in the Ancient Greece. He wrote a document that included several rules on the medical work that came down to us under the so-called Hippocratic Oath, which seems to have been written around the 5th century BC. Even though some of them are no longer accepted, Hippocratic Oath has an important symbolic value and it is common to be known and accepted by new medical graduates in some countries.

Hippocrates established three important rules regarding medical deontology, the principles of beneficence, non-maleficence and confidentiality. The first two were devoted to assure that physicians will do their best to their patients as well they will avoid any damage with their actions. These two principles are still accepted as essential in contemporary medicine and were included in the principlism ethics. The *primum non nocere* dictum that summarizes the second is still often used. The third is an important element of patient-doctor relationship. Hippocratic ethics ignored, however, the principle of patient's autonomy, as physicians had all the power to choose the treatment of their patients in medicine at that time. This traditional behaviour, known as medical paternalism, was maintained during many centuries and was only withdrawn in recent times.

Hippocratic rules were accepted during many years, as it happened with the Galen theory of four humours to explain the pathophysiology of diseases. The influence of Greek medicine also reached Arabic medicine, the most important during the Middle Ages, and with a strong influence in Christian societies of this time. Hippocratic ethics was maintained throughout this time, and a new contribution to the field would have to wait until the 19th century.

Bioethics and the beginning of scientific medicine in the 19th century

The end of 17th century and 18th century saw the establishment of the basis that allowed scientific advances in the next ones. At that time, the use of mathematics, physics and chemistry enabled important biological and medical discoveries. In the ethical field, the main contribution was the book *Medical Ethics* of Sir Thomas Percival (1740–1804), published in 1803, and that was a review of his *Medical Jurisprudence* that appeared in 1794.

In his book, Percival argued that the main function of a physician was to benefit patients by following the Hippocratic rules but again he did not consider the patients' rights to be informed and their consent to be treated. In the area of research, he established the need to consult other colleagues when a physician wanted to try new medical treatments or untested surgical procedures. This was an important new contribution as it advised for the peer review approach, as this recommendation was known later, and was against the possibility of a physician trying a new therapy on his own initiative. Percival's book, written as aphorisms, had a profound impact on his time and was the main inspiration for the development of the *American Medical Association* code of ethics that was adopted in 1847, only one year after its creation (Maehle 2013).

Biomedical research advanced quickly in the 19th century and the need of adopting a respectful approach to human beings was also considered. In this way, the French scientist Claude Bernard (1813–1878), one of the fathers of research in biomedical sciences, recognized in his capital book *Introduction à l'étude de la médecine expérimentale* (1865) that the principle of medical and surgical morality consisted in refraining from performing any experiment on human beings that could harm them in any way, even when the result could be useful for scientific knowledge or the health of other people.

Regardless of these good wishes, in the following years several scientists performed medical experiments that did not follow Bernard's recommendations. For instance, in 1892, Albert Neisser (1855–1916), the discoverer of *Neisseria gonorrhoeae,* tested a syphilitic serum to immunize patients against the disease without giving any information to patients or asking for their consent. In 1873, Gerhard Armauer Hansen (1841–1912), the discoverer of the aetiology of leprosy, injected samples with the causative agent *Mycobacterium leprae* in the eyes of a patient to study the possibility of transmitting the disease, also without the patient's consent. Other well-known cases were the studies of Walter Reed (1851–1902) to establish the participation of mosquitoes in the transmission of yellow fever (1900), or the brain biopsies of patients with tertiary syphilis to confirm the presence of *Treponema pallidum* carried out by Udo Julius Wile (1882–1965) in 1916.

All these questionable research practices had finally some consequences. In 1900, the Prussian government, through the Ministry for Religious, Educational and Medical Affairs, issued a decree that limited the research activities to adult subjects after being informed and when they reported their wishes to participate. However, this law did not consider medical interventions for therapeutic or diagnostic purposes and vaccines. Nevertheless, the law changed after 1931 as a consequence of a study that resulted in the death of 76 neonates in Lübeck after being injected with the BCG vaccine for tuberculosis that was contaminated with Koch's bacilli. As a consequence, the new rules included any kind of medical study, both therapeutic and non-therapeutic.

The Nuremberg trials and the Nuremberg Code

The arrival of the Nazis to the German government in 1933 was followed by a new interpretation of the existing legislation on clinical research. The new authorities considered that the rules did not apply to prisoners in concentration camps. At least 8,000 people were submitted to experimentation in the following years until the end of World War II. None of them was protected by the existing law, as they were not considered as human beings with full rights. They were used to study the effects of the mustard gas, the gangrene, the freezing or the oxygen deprivation, among other awful military experiments. Some kids received petrol injections, were frozen until death or simply killed to perform anatomical dissections (Porter 1999). A detailed analysis of this historical episode can be found in the book edited by Annas and Grodin (1992).

Compared to the Nazi experiments, the Japanese case is less well known. In 1936 the Japanese army created a unit called *Epidemic Prevention and Water Supply Unit*, known as Unit 731. Under the leadership of Shiro Ishii, hundreds of physicians, scientists and technicians worked in this research centre in Ping Fang, a small town in northern Manchuria, then under the Japanese Empire's rule. Unit 731 worked on biological warfare research and the production of bombs with infectious agents for lethal diseases such as anthrax, dysentery, typhus, cholera and bubonic plague. They also carried out ballistics, freezing, high temperature and lethal radiation experiments on Chinese citizens first and then, during the World War II, on war prisoners from the United States, United Kingdom and Commonwealth countries. It has been calculated that bacteriological weapons experiments alone killed 200,000 Chinese (Bourke 2001). After the end of the war, US authorities make an agreement with the Japanese government to obtain the data of the experiments carried out by Unit 731. In turn, the experiments were kept secret and the perpetrators received immunity and were not convicted (Bourke 2001, Porter 1999).

As part of the worldwide concern that the atrocities that were discovered in the Nazi concentration camps would not happen again, in the following years were stablished the United Nations (UN) in 1945, the World Medical Association (WMA) in 1946, and the World Health Organization (WHO) in 1948. After the Nuremberg trials (1946–47), in

1947 the Nuremberg Code established the following ten principles (Porter 1999:651, Annas and Grodin 1992:102–103):

1. The voluntary consent of the human subject is absolutely essential.
2. The experiment should be such as to yield fruitful results for the good of society, unprocurable by other methods or means of study, and not random or unnecessary in nature.
3. The experiment should be so designed as based on the results of animal experimentation and knowledge of the natural history of the disease and other problems under study that the anticipated results will justify the performance of the experiment.
4. The experiment should be so conducted as to avoid all unnecessary physical and mental suffering and injury.
5. No experiment should be conducted where there is an *a priori* reason to believe that death or disabling injury will occur; except, perhaps, where the experimental physicians also serve as subjects.
6. The degree of risk to be taken should never exceed that determined by the humanitarian importance of the problem to be solved by the experiment.
7. Proper preparations should be made and adequate facilities provided to protect the experimental subject again even remote possibilities of injury, disability or death.
8. The experiment should be conducted only by scientifically qualified persons. The highest degree of skill and care should be required through all stages of the experiment of those who conduct or engage in the experiment.
9. During the course of the experiment the human subject should be at liberty to bring the experiment to an end if he has reached the physical or mental state that continuation of the experiment seems to him to be impossible.
10. During the course of the experiment the scientist in charge must be prepared to terminate the experiment at any stage, if he has probable cause to believe, in the exercise of good faith, superior skill and careful judgement required of him that a continuation of the experiment is likely to result in injury, disability, or death to the experimental subject.

One of the most important principles of this Code was the recognition of the informed consent in the research with human beings, and the insistence to avoid any damage to participants. In September 1947, during the first conference of the World Medical Association, medical ethics issues were again protagonists, and a new document was drafted concerning the values of the medical profession. The Geneva Declaration by the World Medical Association, as well as the Universal Declaration of Human Rights by the United Nations, both in 1948, addressed important issues about human dignity and respect for all human beings.

Regardless some criticisms, the Nuremberg Code is considered as a foundational element of informed consent, and was the base of the World Medical Association Declaration of Helsinki (1964). This Declaration also established the differences between the therapeutic research (i.e., with potential benefit for the participant) and non-therapeutic research (i.e., with no potential benefit for the participant but for other people).

The Declaration of Helsinki, as any other international Declaration, did not include sanctions for those who did not comply with its principles. In the following years, some clinical research was still carried out without paying attention to ethical commitments. Some of these studies were reported by Henry K. Beecher (1904–1976), in the United States, and Maurice H. Pappworth (1910–1994), in United Kingdom. However, the most disturbing study was the so-called Tuskegee experiment. It began in 1932 and was carried out by the United States Public Health Service to study the natural history of syphilis at a time where there was a lack of general agreement on the effectiveness of available therapies, basically mercurial and arsenical compounds. The study involved hundreds of African-American syphilitic patients who did not receive penicillin when this drug proved its efficacy to cure the disease in the 1940s. The study continued until 1972, and more than a hundred participants died during the trial. The study was discovered when the press revealed it in 1997. In this decade, US press also reported radiation experiments on humans that were carried out during the cold war years without the consent of those affected.

Many of the unethical behaviours that triggered the birth of bioethics were closely related with clinical research, and will be explained in detail in the *Biomedical research* chapter. However, it is important to mention here that, after the Tuskegee experiment, US government created the *National Commission for the Protection of Human Subjects of Biomedical and Behavioral Research* (1974) and stablished the Belmont Report (1978) with new basic ethical principles: respect for people (autonomy), justice and beneficence. Its applications with informed consent, risk/benefit assessment and the fair selection of research subjects, and also its influence with the two scholars of the National Commission, Beauchamp and Childress, and its consideration of bioethical principles, proposed in 1979, will be addressed in the following chapters.

Bioethics after Nuremberg: the birth of a discipline

Cascais (1997) summarized six aspects that have contributed to the birth of bioethics:

1. Abuses in research on human beings.
2. Availability of new medical technologies on diagnosis and therapeutics.
3. The challenging of prevalent medical paradigms and the ultimate meaning and purpose of medical care.
4. New scientific and social fields of concern dealing with ecology and environmental health, genetic engineering and biotechnologies, demographics, behavioural manipulation or reproductive medicine.
5. The upsurge of social movements raising issues of medical import.
6. The need for an ethics for the technological age and, simultaneously, the seemingly insurmountable impossibility of founding such an ethics as much as an ethics in general.

The birth of modern bioethics as an independent discipline from medical ethics can be established in the 1960s and the 1970s and was favoured by several events (Maehle 2012). First, the development of technology and a scientifically oriented medicine created a gap between physicians and patients. Additionally, theologians, philosophers, lawyers and social scientists became interested in the ethical commitments of medical activities. Social movements of this time led to an empowerment of patients, who wanted to participate in the choices regarding their own health. At the same time, contraceptive techniques, abortion and euthanasia triggered the interest of the Catholic Church.

Biomedical advances of this time challenged the traditional principles of medical ethics, with especial attention to the principle of non-maleficence and inadequate information that limited the principle of autonomy. The availability of haemodialysis for chronic kidney failure created the problem of how to decide patient assignment priorities that exceeded the apparatus available for the procedure at the late 1950s. This situation was a clear ethical conflict that was resolved with the creation of the first ethics committees in hospitals. The first was settled in the Seattle Artificial Kidney Center, and its members were a theologian, a surgeon, a lawyer and a layperson. This committee chose what patients should be treated, and established the criteria for this selection, which included other patient's characteristics other than those strictly medical.

A second challenge was the possibility of performing the first successful kidney transplantations in the 1950s and heart transplantations from 1967. This possibility implied a new dilemma, besides the choice of patients to be treated. The procedure was optimal only if the organs were adequately oxygenated until the transplantation procedure was started. This caused a change in the definition of death in donors that allowed organ harvesting. In 1968, a committee at the Harvard University School of Medicine established the criteria of brain death based on EEG traces. This new concept replaced cardiorespiratory arrest as the traditional criterion for establishing death, and allowed the removal of the organ while it was still receiving an adequate blood supply, since cardiocirculatory function was still maintained.

An important event that also helped to shape the discipline was the creation of two research centres in United States. The first was the Institute of Society, Ethics, and the Life Sciences, in 1969, later known as the Hastings Centre. The second was created at Georgetown University and was called Kennedy Institute of Ethics, in 1971 (Jonsen 2000), that contributed to the publication of two important works the *Encyclopaedia of Bioethics* in 1975 and the *Bibliography of Bioethics*, an annual review that collected the works of the discipline. By the end of the 1970s, bioethics was already considered a new discipline with its own literature and an important number of courses and scientific meetings.

Another aspect that also triggered the bioethical discussions was the consideration of the beginning and the end of life. In the first case, the discussion was established about the abortion procedures after prenatal diagnosis and embryo selection. In the second case, the removal of assisted respiratory support in cases of persistent vegetative states. Some cases had strong debates in the society of the time. One example was the case of Karen Quinlan, a woman in a persistent vegetative state for several months. Her parents asked for the withdrawal of the breathing tube but hospital authorities refused this request. Finally, the New Jersey Supreme Court allowed Quinlan's wishes. Other situations that involved ethical debates were the Roe v. Wade case in 1973, when United States Supreme Court established that a state could not limit woman's right to abortion in the first trimester of pregnancy. Still in 1993, the United Kingdom High Court, Court of Appeal and House of Lords accepted the withdrawal of a feeding tube in the Tony Bland case as legal.

Another important issue that created a continuous debate was the development of medical advances that allowed *in vitro* fecundation, and the birth of Louise Brown, the first child born using such technologies in 1978. From the social point of view, this scientific advance was seen first with an early enthusiasm, later with a full refusal to end with the wide acceptance in contemporary society.

At present, the possibility of selecting embryos free of a genetic disease to avoid the transmission to the potential siblings, and the possibility of using embryos to transplant sick children is still the subject of debate in some social fields. Perhaps, one of the most discussed topics is the possibility of using embryonic stem cells for transplants and therapeutic cloning. In the first case, the technique includes the destruction of embryos in the blastocyte phase that have been obtained in excess in *in vitro* fecundation procedures. This situation poses the question of the possibility of being an abortion procedure and the consequent discussion of this topic.

However, beyond medical dilemmas in the daily practice, bioethics has also a conceptual birth in the 1970s. Even though *Bioethics, the Science of Survival* (Potter 1970) and later *Bioethics: bridge to the future* (Potter 1971) were the first proposals that suggested thinking of bioethics as a global movement to promote concern for ethics, the article *Bioethics as a Discipline* (Callahan 1973) was the first that argued that there must be a new academic discipline beyond medical ethics: "bioethics is not yet a full discipline… (it lacks) general acceptance, disciplinary standards, criteria of excellence and clear pedagogical and evaluative norms" (Callahan 1973:68). This lack or place of vacancy is for Callahan, in fact, the opportunity to define the discipline.

Callahan defines bioethics as a discipline clarifying the "definition of issues, methodological strategies and procedures for decision-making" (Callahan 1973:71) and concludes by saying that the discipline of bioethics "should be so designed, and its practitioners so trained, that it will directly—at whatever cost to disciplinary elegance—serve those physicians and biologists whose positions demand that they make the practical decisions" (Callahan 1973:72).

One look to the 21st century

The importance of bioethics in contemporary times is due to its ability to cope with the complex dilemmas of modern societies, which ultimately help political and social leaders in decision-making. Additionally, the new discipline also helped to save the traditional ethics from a lack of interest in the discipline in the changing world of the late 20th century. Toulmin (1973) suggested that medical ethics and bioethics have prevented traditional ethics from being forgotten by its lack of interest in modern times. In a similar way, Drane (2002:24) clearly explained this point of view: "The problems with which medical ethics grappled not only created a new interest in ethics but also saved ethics from an irrelevance created by an overly abstract, rationalistic, linguistic approach. Philosophers, theologians, lawyers, social scientists suddenly found the ethical aspects of medicine and biosciences to be areas of fascination and started studying and writing about them".

In our times, we consider that any medical issue cannot be fully analysed without considering its ethical dimensions. As an example of this statement, Emanuel et al. (2022) have recently analysed the ethical compromises of COVID-19 pandemic. They discussed the fundamental values that should be considered in a situation of limited medical resources, as it happened at the beginning of the pandemic in First World countries and persisted in those of the Third World. In this case, it was again very important to maximize the benefit of each measure and the prevention of damage, the minimization of disadvantages, the instrumental value and the avoiding of discrimination for any reason. The authors finished their analysis by stating: "Ethics must be integrated into emergency decision making. Ethicists should be at the table when policies are formulated, rather than merely serving as external critics." This position was very different from what happened in AIDS, the previous world pandemic where, for instance, the criminalization of some social groups was the rule.

However, nowadays bioethics also addresses problems beyond medical attention. As Callahan (2015) says, as sociological, anthropological and psychological studies proliferated, bioethics interest turned towards health policy and the use of scientific knowledge in public policies. The application of bioethics to global problems of human health and survival such as nuclear war and global warming concerns worldwide.

In 1988, Van Rensselaer Potter proposed the concept "global bioethics", as he became concerned with globalization, world population growth, economic and gender inequalities, and environmental change as problems for bioethics. However, the definition of global bioethics is not unified. It can be understood that it is an attempt to globalize the concerns of bioethics by focusing more attention on the problems of countries with scarce resources, public health, global justice and equity, and also a statement about the correct way to practice bioethics and that bioethics becomes a global research field. According to Henk Ten Have (2016), the greatest foundation of global bioethics is relationality and it is necessary to redefine individual autonomy as "relational autonomy".

All these last proposals, finally, can be understood as a broad way of understanding the bioethics discipline, with the inspiration proposed by Fritz Jahr in 1927, that is: the responsibility of human beings with their fellows, animals and plants, that is, with the whole planet.

References cited

Annas, G.J. and M.A. Grodin. 1992. The Nazi Doctors and the Nuremberg Code. Oxford University Press, New York.

Baker, R.B. and L.B. McCullough (eds.). 2009. The Cambridge World History of Medical Ethics. Cambridge University Press, New York.

Beauchamp, T.L. and J.F. Childress. 2019. Principles of Biomedical Ethics. Oxford University Press, New York.

Beecher, H.K. 1996. Ethics and clinical research. N. Eng. J. Med. 274: 1354–1360.

Bernard, C. 1957. Introduction to the Study of Experimental Medicine. Dover, New York.

Bourke, J. 2001. The Second World War: A People's History. Oxford University Press, New York.
Callahan, D. 1973. Bioethics as a discipline. Hastings Center Studies 1: 66–73.
Callahan, D. 2015. Bioethics: Its Past and Future. Global Bioethics: What For? UNESCO, Paris.
Cascais, A.F. 1997. Bioethics: History, scope, object. Global Bioethics 10: 9–24.
Chadwick, R. and D. Wilson. 2018. The emergence and development of bioethics in the UK. Medical Law Review 26(2): 183–201.
Chiong, W. 2020. Insiders and outsiders: Lessons for neuroethics from the history of bioethics. AJOB Neuroscience 11(3): 155–166.
Dhai, A. 2014. The research ethics evolution: from Nuremberg to Helsinki. S. Afr. Med. J. 104(3): 178–180.
Drane, J.F. 2002. What is bioethics? A history. pp. 15–31. *In*: Losas, F. and L. Agar (eds.). Interfaces of Bioethics and the Social Science. Third Meeting of the International Advisory Board on Bioethics. Pan American Health Organization. World Health Organization.
Emanuel, E.J., R.E.G. Upshur and M.J. Smith. 2022. What Covid has taught the world about ethics. N. Eng. J. Med. 387(17): 1542–1545.
Faunde, T.A. 2005. Will international human rights subsume medical ethics? Intersections in the UNESCO Universal Bioethics Declaration. J. Med. Ethics 31: 173–178.
Giovanni, R. 2001. Potter's personal history of bioethics. An examination and survey. Global Bioethics 14(4): 63–71.
Heiman, T. 1917. Etyka lekarska i abowiqzki lekarza (deontologia). Gebethner i Wolff, Warszawa.
Jahr, F. 1926. Wissenschaft vom Leben und Sittenlehre (Alte Erkenntnisse in neuem Gewande). Mittelschule. 40(45): 604–605.
Jahr, F. 1927. Bio-Ethik: eine Umschau über die ethischen Beziehungen des Menschen zu Tier und Pflanze. Kosmos. 24(1): 2–4.
Jonsen, A.R. 1998. The Birth of Bioethics. Oxford University Press, New York.
Jonsen, A.R. 2000. A Short History of Medical Ethics. Oxford University Press, New York.
Jonsen, A.R. 2004. The history of bioethics. pp. 31–51. *In*: Khushf, G. (ed.). Handbook of Bioethics: Taking Stock of the Field from a Philosophical Perspective. Kluwer, New York.
Lerner, B.H. and A.L. Caplan. 2016. Judging the past: how history should inform bioethics. Ann. Intern. Med. 164(8): 553–557.
Lima, N. 2009. Fritz Jahr y el Zeitgeist de la Bioética. Aesthethika. 5: 4–11.
Lolas Stepke, F. 2008. Bioethics and animal research. A personal perspective and a note on the contribution of Fritz Jahr. Biol. Res. 41: 119–123.
Löther, R. 1997. Evolution der Biosphäre und Ethik. pp. 61–68. *In*: Engels, E.M., T. Junker and M. Wiengarten (eds.). Ethik der Viowissenschaften: Geschichte und Theorie – Beiträge zur 6. Jahrestagung der Deustchen Gesellschaft für Geschichte und Theorie der Biologie (DGGTB) in Tübingen. Verlag für Wissenschaft und Bildung, Berlin.
Maehle, A.H. 2013. Medical ethics and the law. pp. 543–560. *In*: Jackson, M. (ed.). The Oxford Handbook of the History of Medicine. Oxford University Press, New York.
Moreno, J.D., U. Schmidt and S. Joffe. 2017. The Nuremberg Code 70 years later. JAMA 318(9): 795–796.
Muzur, A. and I. Rincic. 2015. Two kinds of globability: a comparison of Fritz Jahr and Van Rensselaer Potter's bioethics. Global Bioethics 26(1): 23–27.
Pessini, L. 2013. At the origins of bioethics: from Potter's bioethical creed to Fritz Jahr's bioethical imperative. Rev. Bioét. 21(1): 9–18.
Porter, R. 1999. The Greatest Benefit to Mankind. A Medical History of Humanity. W. W. Norton & Company, New York.

Potter, V.R. 1970. Bioethics, the science of survival. Perspectives in Biology and Medicine 14: 127–153.
Potter, V.R. 1971. Bioethics: Bridge to the Future. Prentice Hall, New Jersey.
Potter, V.R. 1988. Global Bioethics: Building on the Leopold Legacy. MSU Press, East Lansing, MI.
Sass, H.M. 2007. Fritz Jahr's bioethischer Imperativ. 80 Jahre Bioethick in Deutschland von 1927 bis 2007. Zentrum für medizinische Ethik, Bochum.
Sass, H.M. 2008. Fritz Jahr's 1927 concept of bioethics. Kennedy Institute of Ethics Journal 17(4): 279–295.
Stark, L. 2016. The unintended ethics of Henry K. Beecher. Lancet. 387: 2374–2375.
Ten Have, H. 2016. Global Bioethics. An Introduction. Routledge, New York.
Toulmin, S. 1973. How medicine saved the life of ethics. Perspectives in Biology and Medicine 25(4): 736–50.

Chapter 3

The Principlist Approach in Bioethics

Ester Busquets-Alibés[a,*] and *Lydia Feito*[b]

Introduction

The principles approach of T.L. Beauchamp and J.F. Childress, formulated in 1979 in their book *Principles of Biomedical Ethics*[1] (PBE) has had an enormous influence on the development of bioethics, to the point that for a long time it was a core element that people worked with or argued against. As R. Gillon (1994) and D. Gracia (1997) stated, principlist model was always present and, deep down, all the foundational theories of bioethics had to necessarily dialogue with it. If, more than forty years later, Beauchamp and Childress' four principles framework continues to play a significant role in the evolution and maturation of bioethics, it is, above all, because the authors themselves have taken criticism seriously and have attempted to revise and reformulate their theoretical approach to make it more robust. As J.J. Ferrer (2016:114) points out, "the formulation of alternative approaches [to principlist model] shows its insufficiency", but at the same time "the axiological poles identified by Beauchamp and Childress are undeniable reference points for bioethical debates".

The principles of bioethics appear as the North American attempt to provide the young discipline of bioethics with moral principles that could have general applicability. In 1974 the United States Congress, aware of

[a] Universitat de Vic – Universitat Central de Catalunya, Spain.
[b] Universidad Complutense de Madrid, Spain.
* Corresponding author: ester.busquets@uvic.cat

[1] In this chapter we work with the eighth edition of the book *Principles of Biomedical Ethics*, published in 2019.

abuses carried out in research with human beings, commissioned the National Commission for the Protection of Human Subjects of Biomedical and Behavioral Research to carry out an extensive investigation that would identify fundamental ethical principles to guide scientific research, as well as the development of specific guidelines to ensure that research is carried out in accordance with these principles. The identification of general ethical principles was a novel task. The National Commission worked between 1975 and 1978, and in 1978 *The Belmont Report: Ethical Principles and Guidelines for the Protection of Human Subjects of Research* was published, a document that has become one of the founding texts of bioethics. The Belmont Report proposal includes three principles: respect for people, beneficence and justice.

The great problem that the commission encountered was how to go beyond the codes that, while useful, were not operative to resolve the conflicts that arose in research with human beings, since its rules were perceived as inadequate in complex situations. They therefore suggested that the use of rules that were difficult to interpret and apply in practice should be replaced by another method based on the acceptance of a series of ethical principles that would lay the foundations for a subsequent formulation of specific rules. Thus, practical procedures follow from each of the principles: from the principle of autonomy (respect for people) follows the procedure of informed consent; from the principle of beneficence follows the evaluation of risks and benefits; and from the principle of justice follows the equitable selection of subjects of experimentation. This is, in the opinion of Gracia (1991), the true merit of bioethics that reaches a new methodological approach that would be highly influential.

Beauchamp and Childress attempted to make these principles applicable to a broader field than human research. Their approach, also conceived in the same year as the Belmont Report, was published in 1979. One of the authors, Beauchamp, had been part of the National Commission and was well acquainted with its deliberations. Moreover, for some years they had been reflecting on the need for principles for ethical decisions in medicine, which is the objective they set out in the book's preface: to systematically analyse the moral principles that should be applied in biomedicine.

It is important to note that the *Belmont Report* had made no effort to support with arguments the fundamental convictions linked to the principles. According to Ferrer (2003), they are simply pronounced, assuming them to be valid and turning them into the cornerstone of the whole conceptual edifice that was to be built. It should be noted that Beauchamp and Childress developed and deepened the fundamental intuition of the *Belmont Report*, and for this reason they made the effort—throughout the eight editions—to try to provide a theoretical base for their approach.

Ethical foundation

A moral principle is a normative statement that establishes what should be done to protect a value. It is therefore a methodological element for decision-making. The principled approaches in ethics have a marked deontological accent; that is, they establish an *a priori* guideline for action so that we know what we must do because it is in accordance with the safeguarding of values that we consider important. They thus create a way of understanding ethics that is very different from consequentialist approaches, where the goodness of the action is determined by its consequences, thus constituting an *a posteriori* analysis.

Beauchamp and Childress defend different philosophical convictions. Childress is a deontologist, more akin to these models of *a priori* principles, while Beauchamp is situated in the rule utilitarianism field, a form of consequentialism that establishes that actions are correct to the extent that they conform to a rule, whose ultimate goal must be to achieve the greatest good. The authors consider that this difference in their approaches, far from being insurmountable, is an advantage as it ensures that theoretical discrepancies do not prevent agreement on rules and procedures. Both accept a system of principles that serves to make decisions in specific cases.

As will be said later, Beauchamp and Childress consider that their principles are *prima facie*; that is, they are not real and effective duties—using the distinction defended by W.D. Ross in 1930 (Ross 1994). In order to be applied in real life, they have to undergo a specification process that allows the establishment of concrete rules for the particular case. This means that principles are always binding and have no exceptions, but the rules that are applied to the cases can have exceptions. This procedure seems infallible and overcomes the apparent difficulty of universal and absolute principles. However, it forces them to consider all the principles to be of equal rank, without it being possible to place them in a hierarchy, unless there is a conflict between them, in which case, depending on the rule applicable to the case, the rule that takes precedence is determined. For many, this weakens the deontology of the approach (Gracia 1991), which is why its critics consider that the North American current of bioethics initiated by Beauchamp and Childress is dominated by rule utilitarianism. On the other hand, it brings the approach closer to the casuistry perspective defended by two other authors who also participated in the National Commission: Jonsen and Toulmin (1988). From this perspective, which has been proposed as an alternative to principlist model, the novelty of the Belmont Report is considered to lie in introducing a casuistic approach, where the principles are understood as maxims that help to resolve conflicts and, therefore, has a procedural vision.

Common morality

In the seventh edition of PBE Beauchamp and Childress, pushed by criticisms from the philosopher Bernard Gert (Gert et al. 2006), incorporated the concept of common morality. The theory of common morality holds, fundamentally, that "some core tenets found in every acceptable particular morality are not relative to cultures, groups, or individuals. All persons living a moral life know and accept rules such as not to lie, not to steal others' property, not to punish innocent persons, not to kill or cause harm to others, to keep promises, and to respect the rights of others" (Beauchamp and Childress 2019:3). The violation of these norms is contrary to ethics and generates both remorse in the moral agent and censorship in the community.

According to Beauchamp and Childress, "the set of universal norms shared by all persons committed to morality is the common morality. This morality is not merely a morality, in contrast to other moralities. It is applicable to all persons in all places, and we appropriately judge all human conduct by its standards" (2019:3). The imperatives derived from common morality are: "(1) Do not kill, (2) Do not cause pain or suffering to others, (3) Prevent evil or harm from occurring, (4) Rescue persons in danger, (5) Tell the truth, (6) Nurture the young and dependent, (7) Keep your promises, (8) Do not steal, (9) Do not punish the innocent, and (10) Obey just laws" (Beauchamp and Childress 2019:3).

The common morality also contains standards other than obligatory rules of conduct. Some moral character traits, or virtues, have to be developed, such as: "(1) nonmalevolence (not harboring ill will toward others), (2) honesty, (3) integrity, (4) conscientiousness, (5) trustworthiness, (6) fidelity, (7) gratitude, (8) truthfulness, (9) lovingness, and (10) kindness. These virtues are universally admired traits of character … In addition to the obligations and virtues … the common morality supports human rights and endorses *moral ideals* such as charity and generosity"[2] (Beauchamp and Childress 2019:3–4).

The general principles proposed in PBE are not derived from any specific moral theory, but are rather presented as principles derived from the common moral heritage of humanity. Basically, it is a theory that is very close to the tradition of natural moral law. Critics of the theory of common morality claim that the historical or anthropological evidence for its existence is inconsistent (Christen et al. 2014). Although it is not possible to question the convenience of some moral minimums that guarantee

[2] The authors of PBE, after dedicating the first chapter to normative ethics, in the second chapter focus on virtue ethics and extensively develop the five focal virtues for health professionals: compassion, discernment, trustworthiness, integrity, and conscientiousness (Beauchamp and Childress 2019:31–64).

citizen coexistence, it is not clear that the common morality proposed by Beauchamp and Childress has the universal acceptance they claim.

The four principles they expounded in PBE are only a small part of the universal common morality. They do not pretend to make a catalogue of the basic principles of common morality, but rather extract these four principles from the common morality to build a normative framework for biomedical ethics (Beauchamp 2020).

Reflective equilibrium and common morality

For Beauchamp and Childress, the theoretical justification of the principles does not depend solely on their being rooted in common morality, but rather should be submitted to the test of reflective equilibrium. Reflective equilibrium, a term coined by John Rawls (1971), consists in finding a consonance between the various elements of a moral system: principles, judgements and underlying theories. The goal is the search for coherence between the different elements of moral reflection. A theory or a set of moral beliefs would be justified if they are accepted on the basis of the reflective examination that has been outlined.

Moral reflection is based on a set of beliefs that are considered acceptable, without yet having an argued justification. This is what Rawls called considered judgments, moral convictions that we trust and that we think are less influenced by prejudices and partiality. "Whenever some normative feature in a person's or a group's prevailing structure of moral views conflicts with one or more of their considered judgments, they must modify something in their viewpoint and strive to achieve equilibrium and overall coherence. Even the considered judgments that we accept as central in the web of our moral beliefs are subject to revision once a conflict occurs" (Beauchamp and Childress 2019:440).

It is important to note that coherence alone is insufficient to morally justify an action or a practice. Coherence could be the result of a set of prejudices without moral foundation. That is why reflective equilibrium must be in tandem with common morality. As we have said, ethical reflection begins with the set of weighted judgements, coming from common morality, which cannot be justified in terms of other more fundamental moral principles under penalty of falling into an infinite regress or vicious circle. The method of reflective equilibrium is applied once the weighted judgements that constitute the starting point of all ethical reflection have been established; or, put differently, one starts from the principles—weighted judgements—based on common morality and submits them to the method of reflective equilibrium.

The content of the four principles of bioethics

Beauchamp and Childress (2019) argue that the vision of the four moral principles is the best framework for bioethics:

1. Respect for autonomy (a principle that requires respect for decisions and the ability of autonomous people to make decisions),
2. Nonmaleficence (a principle that requires avoiding causing harm to others),
3. Beneficence (a principle that requires reducing or preventing harm, as well as helping others, balancing benefits, burdens and risks), and
4. Justice (a principle that requires a fair distribution of benefits and burdens among all affected parties).

Respect for autonomy

Beauchamp and Childress (2019) have often been accused of giving preference to the principle of autonomy. For this reason, they point out that the fact that the analysis of the principles begins with the respect for autonomy does not mean that this principle has priority over the others, and at the same time they maintain that respect for autonomy is not an excessively individualist, rationalistic and legalist principle. They argue that autonomous persons are those who act freely in accordance with the course of action that they themselves have chosen. In contrast, a person with diminished autonomy is substantially controlled by others or is prevented for some reason from deliberating or acting on their wishes or plans.

The interest of Beauchamp and Childress consists in being able to determine when a human action can be considered autonomous, and establishing the "theory of the three conditions" (2019:103): intentionality, understanding and the absence of external controls. For an action to be autonomous, the moral agent needs to have the intention of doing what he/she is doing, to understand the action and to be free from external influences that can control it. They argue that the first condition does not admit degrees, it is either intentional or not intentional, whereas understanding and the absence of controls do admit degrees. The autonomous actions of people impose a double duty on others: one negative and one positive. Although the negative duty is not absolute, it prohibits coercive intervention so that people cannot act according to their autonomous options. The positive duty requires that both a correct process of information and all the actions necessary to help people act autonomously be carried out. The exercise of autonomy is often not feasible without the cooperation of other people. Beauchamp and Childress point out that respect for autonomy is the foundation of numerous moral truths, such as "telling the truth, respecting the privacy of others, protecting confidential information and obtaining consent for interventions with patients" (2019:105). Positive duties also require specifications that give rise to exceptions in their application. For example, in emergency situations it is not necessary to comply with all the requirements of informed consent.

Beauchamp and Childress maintain that "obligations to respect autonomy do not extend to persons who cannot act in a sufficiently autonomous manner and to those who cannot be rendered autonomous because they are immature, incapacitated, ignorant, coerced, exploited, or the like. Infants, irrationally suicidal individuals, and drug-dependent patients are examples" (2019:105–106). When people have diminished autonomy and it is not possible to get them to decide autonomously; then, and only then, can interventions of a paternalistic nature be justified.

The practical application of the principle of respect for autonomy, as we have already said, is informed consent, the elements of which are: "(1) competence (capacity or ability), (2) disclosure, (3) understanding (comprehension), (4) voluntariness (in deciding), and (5) consent (decision in favor of a plan and authorization of the chosen plan)" (Beauchamp and Childress 2019:122–123).

Nonmaleficence

Beauchamp and Childress (2019), unlike the Belmont Report and the famous work of William Frankena (1963), distinguish between the principle of beneficence and the principle of nonmaleficence. According to them, nonmaleficence, inspired in the classical maxim *Primum non nocere,* is defined as the obligation that one ought not to inflict evil or harm. In this regard, the obligations of nonmaleficence require that the moral agent refrain from carrying out certain actions. The obligations of nonmaleficence are negative and, in general, are more demanding than the positive obligations of the principle of beneficence. We must remember that the principles of Beauchamp and Childress (2019) are *prima facie,* therefore we cannot state that the norms of nonmaleficence always have priority over those of beneficence.

In PBE harm is defined as frustrating, defeating or delaying someone's interests without necessarily constituting an injury or injustice to the injured party. Although harmful actions are *prima facie* wrong, they can be justified in certain circumstances. Moral obligation prohibits unjustified actions, but if a harm has proportional justification, it can be accepted. Beauchamp and Childress (2019) give as an example the amputation of a gangrenous limb, which is a harm intended to achieve a greater good. The book explores "the principle of nonmaleficence and its implications for several areas of biomedical ethics where harm may occur distinctions between killing and allowing to die, intending and foreseeing harmful outcomes, withholding and withdrawing life-sustaining treatments, as well as controversies about the permissibility of physicians assisting seriously ill patients in bringing about their deaths" (Beauchamp and Childress 2019:155).

Like all the principles, nonmaleficence gives rise to a broad set of more specific moral rules. Although there are many rules, Beauchamp and Childress list five of them: "(1) Do not kill. (2) Do not cause pain or suffering. (3) Do not incapacitate. (4) Do not cause offense. (5) Do not deprive others of the goods of life. Both the principle of nonmaleficence and its specifications into these moral rules are *prima facie* binding, not absolute" (2019:159).

Beneficence

Morality is not limited to respect for autonomy and nonmaleficence but also includes contributing to the well-being of others. "Principles of beneficence potentially demand more than the principle of nonmaleficence because agents must take positive steps to help others, not merely refrain from harmful acts. An implicit assumption of beneficence undergirds all medical and health care professions and their institutional settings" (Beauchamp and Childress 2019:217).

Beauchamp and Childress "examine two principles of beneficence: positive beneficence and utility. The principle of positive beneficence refers to a statement of a general moral obligation to act for the benefit of others. The principle of utility requires agents to balance benefits, risks, and costs to produce the best overall results. The principle of utility in our account is therefore not identical to the classic utilitarian principle of utility. Whereas utilitarians view utility as a fundamental, absolute principle of ethics, we treat it as one among a number of equally important *prima facie* principles. The principle of utility we defend is legitimately overridden by other moral principles in a variety of circumstances, and likewise it can override other *prima facie* principles under various conditions" (2019:218).

Beauchamp and Childress are aware that there are those that think that beneficence is a moral idea, and therefore there can be no duties of beneficence. They respond to this objection by arguing that beneficence does not require acts of severe sacrifice and extreme altruism, but that does not nullify the strict duties of beneficence. For this reason, they introduce the distinction between specific and general beneficence. The former extends to people with special ties (family, friendship or professional ties, among others), and the latter is supererogatory, it goes beyond the call of duty, but it can be imposed if certain conditions are met, such as when there is risk to the life, health or other fundamental interest of a person, or the action is necessary to prevent harm.

The thesis that beneficence expresses a primary obligation in health care is ancient. But when beneficence is exercised in violation of respect for the beneficiary's autonomy, it leads to paternalism. An act is paternalistic when a person's preferences and decisions are exceeded, cancelling or limiting their autonomy for the good of the person whose autonomy has been violated. Paternalistic interventions are only justified in some circumstances, for example in emergency situations in a hospital.

The practical application of the beneficence principle consists of "balancing benefits, costs and risks. The authors examine formal techniques of analysis (risk-benefit analysis (RBA), cost-benefit analysis (CBA) and cost-effectiveness analysis (CEA)) and conclude that, with suitable qualifications, there are morally unobjectionable ways to explicate the principle of utility, as a principle of beneficence, but that principles of respect for autonomy and justice often should be used to set limits on the uses of these techniques" (Beauchamp and Childress 2019:257).

Justice

Justice has to do with what is due to people, with what in some way belongs or corresponds to them. In the biomedical field, the most interest dimension of justice is distributive justice, which refers to the equitable distribution of rights, benefits, responsibilities and burdens in society. To determine whether the distribution of burdens and benefits is just, we need to turn to criteria of justice for guidance in this distribution. These criteria can be formal or material (Beauchamp and Childress 2019:267–268). The so-called principle of formal justice, traditionally attributed to Aristotle, is defined as "equals must be treated equally, and unequals must be treated unequally". It is a formal principle because it lacks concrete content. It does not tell us from what point of view the cases are or should be equal, nor does it provide us with criteria to determine equality. It simply tells us that people who are equal should be treated equally. In the absence of concrete content, we need to have material criteria of distributive justice.

Material criteria refers to the concrete contents that fill the empty structure of the formal precept. Some material principles of distributive justice that are found in traditional ethical theories are: "(1) To each person according to rules and actions that maximise social utility (utilitarianism). (2) To each person a maximum of liberty and property resulting from the exercise of liberty rights and participation in fair free-market exchanges (libertarianism). (3) To each person according to principles of fair distribution derived from conceptions of the good developed in moral communities (communitarianism). (4) To each person an equal measure of liberty and equal access to the goods in life that every rational person values (egalitarianism). (5) To each person the means necessary for the exercise of capabilities essential for a flourishing life (capability theories). (6) To each person the means necessary for the realisation of core elements of well-being (well-being theories)" (Beauchamp and Childress 2019:270–271).

"No obvious barrier prevents acceptance of more than one of these principles as valid—perhaps all six—in a pluralistic theory of justice. However, these principles are considered competitive in much of the literature on general theories of justice. To retain all six, one would have to argue that each of these material principles identifies a *prima facie* obligation whose weight cannot be assessed independently of particular goods and domains in which they are applicable, and then one would have to show these principles can be rendered coherent in a pluralistic theory of justice" (Beauchamp and Childress 2019:271).

It should be said that in any society different criteria of justice are used in different areas of life. According to Beauchamp and Childress, "the polices of just access to health care, strategies of efficiency in health care institutions, and global needs for the reduction of health-impairing conditions dwarf in social importance every other issue considered in PBE. Global justice and just national health care systems are distant goals for the many millions of individuals who encounter barriers to access health care and to better health. Every society must ration its resources, but many societies can close gaps in fair rationing more conscientiously than they have to date. They propose recognition of global rights to health and enforceable rights to a decent minimum of health care in nation-states, while recognising that adequately securing these rights in political states and globally is an exceedingly ambitious and difficult path to pursue, even when the goals of policy are strongly supported by principles and theories of justice" (Beauchamp and Childress 2019:313–314).

Specification and weighting of the four principles

The four clusters of principles we present do not by themselves constitute a general ethical theory. They provide only a framework of norms with which to get started in biomedical ethics. These principles must be specified in order to achieve more concrete guidance.

Beauchamp and Childress, inspired by the work of William David Ross (2002), consider the four principles to be *prima facie* duties. A *prima facie* obligation binds us, unless it conflicts with an obligation at the same or higher level. This means that in the case of a conflict between duties, the moral subject has to think carefully in order to establish what the actual duty is in the concrete circumstances of the moment. This weighting up forces the subject to assess the different courses of action and choose that which enables him/her to obtain the greatest balance of right over wrong (Beauchamp and Childress 2019). However, before weighting the *prima facie* obligations, these principles need to be specified.

Specifying principles and rules

The four principles are very general and, by themselves, cannot guide us in making moral decisions. That is why they need to be specified so as to give us concrete guidelines for the moral life. Specification is a process that consists of descending from the general to the specific, thus indicating which situations are governed by the general principle or rule. "Specification is a process of reducing the indeterminacy of abstract norms and generating rules with action-guiding content. For example, without further specification, 'do no harm' is too bare for thinking through problems such as whether it is permissible to hasten the death of a terminally ill patient. Specification is not a process of producing or defending general norms such as those in the common morality; it assumes that the relevant general norms are available" (Beauchamp and Childress 2019:15).

Specification is not without controversy when moving from general principles to concrete rules because we can agree with moral obligation to "do no harm" but disagree about the specificity in a particular case. Specification does not seek to justify or explain the content of the rules but to specify their scope of application in the specific situation, in order to identify what should be done or avoided. Specification adds content to general rules. "All moral rules are, in principle, subject to specification. All will need additional content because, as Richardson (2000) puts it, 'the complexity of the moral phenomena always outruns our ability to capture them in general norms'. Many already specified rules will need further specification to handle new circumstances of conflict" (Beauchamp and Childress 2019:19).

"To say that problem or conflict is resolved or dissolved by specification is to say that norms have been made sufficiently determinate in content that, when cases fall under them, we know what must be done. Obviously some proposed specifications will fail to provide the most adequate or justified resolution. When competing specifications emerge, the proposed specifications should be based on deliberate processes of reasoning. Specification as a method can be connected to a model of justification that will support some specifications and not others" (ibid.).

Weighting and balancing

For Beauchamp and Childress, specification cannot be confused with weighing, which is a useful and necessary process for the proper resolution of conflict situations. "Balancing occurs in the process of reasoning about which moral norms should prevail when two or more of them come into conflict. Balancing is concerned with the relative weights and strengths of different moral norms, whereas specification is concerned primarily with their range and scope, that is, their reach when narrowing the scope of pre-existing general norms. Balancing consists of deliberation and judgment about these weights and strengths. It is well suited for reaching judgments in particular cases, whereas specification is especially useful for developing more specific policies from already accepted general norms" (Beauchamp and Childress 2019:20).

One of the objections that have been made to weighting is that it involves a process that it is too intuitive and subjective. Beauchamp and Childress consider that weighting must be accompanied by good reasons. Although there are situations of intuitive weighting, the process should not be based solely on intuition and feelings, but must be subjected to rational analysis. To avoid excessive intuition in weighting, Beauchamp and Childress propose conditions that limit the partiality and arbitrariness of the process, helping it to become a rigorous development of reasoned deliberation: "(1) Good reasons are offered to act on the overriding norm rather than the infringed norm. (2) The moral objective justifying the infringement has a realistic prospect of achievement. (3) No morally preferable alternative actions are available. (4) The lowest level of infringement, commensurate with achieving the primary goal of the action, has been selected. (5) All negative effects of the infringement have been minimized. (6) All affected parties have been treated impartially" (Beauchamp and Childress 2019:23).

Criticisms and alternative proposals to principlist model

The four principles approach of Beauchamp and Childress is not without controversy. The truth is that its influence in bioethics has been enormous, to the point that for a long time it was a core element that people worked with or argued against, but as Gillon (1994) stated, it was always present. Its period of greatest influence was after its publication and during the 80s and 90s. As Pellegrino (1994) highlighted, the training of doctors with the principlist doctrine for many years radically changed the way of understanding medical ethics by providing a valid mechanism for decision-making. And there were also extensive discussions about its suitability. One of the most important and favourable reviews of the theory of the four principles was presented in the volume *Principles of Health Care Ethics*, edited by Gillon and Lloyd (1994); another, less favourable, was published in *A Matter of Principles? Ferment in U.S. Bioethics* (DuBose et al. 1994).

In the 90s, the strongest criticisms began to be voiced. Of these, it is worth highlighting those formulated by Clouser and Gert (1990)—in a famous article in a monograph dedicated to the criticism of bioethics in the *Journal of Medicine and Philosophy*—who accused the model of "principlism", that is, of a certain form of reductionism in its approach, a mantra—as Jonsen (1991) would say—that was repeated automatically and without solid moral foundation. In their opinion, the principles were useless as guides to action, they only acted as a list of considerations when discussing bioethical problems. They argued that the principles had no connection with each other and that they easily conflicted, without there being any way of articulating them or an underlying ethical theory that gave them systematic unity. In their opinion, "the principles function neither as adequate surrogates for moral theories nor as directives or guides for determining the morally correct action" (Clouser and Gert 1990:221).

In a subsequent publication, Gert et al. (1997) persisted in their critique, arguing that the principlist movement was disseminated as a generalist model for everyone that really has limitations in its application to practical cases. And, in their opinion, despite the changes that have been introduced in successive editions, these problems have not been resolved.

Also, Pellegrino and Thomasma (1993) criticise principlism based on their theory of virtue. From their perspective, the principlist model is far from clinical reality and the personal history of patients. In addition, they consider that the primacy of the principle of autonomy shows an American and liberal origin of the principles. The accusation of the pre-eminence of the principle of autonomy is one of the most frequent. It can also be seen in the criticism of Holm (1995) who, in his analysis of the fourth edition of PBE, assesses the weaknesses of the principles that he considers to be strongly influenced by a selfish individualism, by an American liberal vision that has little to do with the sense in which human relations are understood in Europe and other parts of the world. Holm concludes that the principles can help to analyse moral problems as long as they are reworked and completed in each cultural context, and their error lies precisely in having been proposed as a universal moral theory.

Beauchamp (2020) responds to this criticism by arguing that it lacks justification because, first, the principle of autonomy is not assumed to be pre-eminent, and second, the universal rules they refer to have nothing to do with the culture of the United States but rather appeal to a universal morality that is related to the basic claims supported by human rights.

With this argument he also tries to answer the criticism of those who have attempted to develop an alternative proposal for European bioethical principles. Kemp and Rentdorff (2000) led a project to develop principles that are more akin to European culture. Their approach is quite different from that presented by Beauchamp and Childress, appealing to values that they consider more important in this other cultural context. They also defend four principles—respect for autonomy, dignity, integrity and vulnerability—but the principles seem more like virtues or values, without the normative content of principles. One could even say, as Valdés (2019) points out, that they are ontological conditions of human beings, rather than principles.

Nevertheless, many criticisms have been made denouncing the influence of North American culture on the principles, with its emphasis on the ideal of autonomy. For many, the reference to a common morality, devoid of cultural influences, is not so clear. There are thus the criticisms made by Campbell (2000), who considers that the principles of Beauchamp and Childress show a certain colonialism, by Neves (2009), who argues that the pre-eminence of autonomy is an approach foreign to European culture, and by Tealdi (2005), who contends that the principles were intended to serve pragmatically for the resolution of clinical cases, without success.

Furthermore, despite the insistence of Beauchamp and Childress on the need to specify the principles in order to convert them into useful moral norms, taking into account what can be interpreted as an overlapping consensus in the manner of J. Rawls,[3] the truth is that the resolution of conflicts in the event of a clash between principles is mostly resolved casuistically, which is why it has been interpreted as, despite its deontological approach, ultimately falling into a form of consequentialist utilitarianism. To overcome this difficulty, Gracia (1989) proposed a form of hierarchical principlism in which two levels were established that were, in his opinion, more in line with the European approach: one of minimums, that contains the principles of nonmaleficence and justice—that would be basic to safeguard the life of individuals and societies—thus placing the social dimension of equity and solidarity first; and another of maximums, in which the principles of beneficence and autonomy are located—necessarily related so as to be able to attend to the preferences of the patient without leading to paternalism.

Despite the interest of this perspective, the North American approach, centred on the pre-eminence of autonomy, has not accepted it. Subsequently, the shortcomings of the principlist approach have led the same author (Gracia 2011) to change his perspective to a deliberative approach in which the principles are abandoned in favour of an articulation of discrepant values, in accordance with the Aristotelian prudential model. This is a sign that the validity of the principles no longer holds and that, today, despite still being enormously important and a necessary reference point, the principlist approach is seen as renewable and more specific to the North American context.

---

[3] According to Rawls, an overlapping consensus on the principles of justice can be produced despite considerable differences in citizen's conceptions of justice.

Conclusion

The principlist approach has proven its strength over the years. Despite the criticisms, the enormous influence that the principles have had in bioethics is undeniable. Its popularity is so great that even those who barely know this discipline have heard of the principles and consider them a fundamental contribution to bioethics. It is no surprise that the four principles of Beauchamp and Childress refer to fundamental values that are inscribed in and legitimise clinical practice, and, as we have seen, also research with human beings: promoting the well-being of patients, recognising their autonomy and making decisions within a framework of social justice are essential elements for the analysis of the problems that may arise in a healthcare setting.

Over several decades other principles have been proposed, mostly to complete the four classical ones. Thus, the principle of truthfulness, the precautionary principle, the principles of European bioethics and so forth have been discussed. All these proposals have dialogued with the traditional approach, emphasising aspects that had not been very visible or that needed complementation. However, it should be said that, although they may be supplemented or placed into dialogue with other approaches, the four principles are still valid.

The most interesting aspect of this finding lies in the fact that, despite the need to attend to the specific reality of particular cases and to analyse the consequences derived from the actions, it is clear that an approach based on *a priori* principles that guide our decisions is the most adequate one. The principles account for how we believe things "should be", serve as a normative framework to safeguard fundamental values and, therefore, are supported by a robust foundation, which protects us against arbitrariness and dogmatism.

The principles speak of elements that are considered substantial; such is the position that Beauchamp and Childress have defended by emphasising their roots in a common morality. However, there is a serious difficulty in accepting this general approach that seems to insist on the existence of material content common to all humanity. In reality, as occurred with the proposal of principles for universal bioethics addressed by UNESCO in 2005 (UNESCO 2006), principles of a common morality cannot but be limited to formal principles given that agreement on specific content requires a cultural, social and political contextualisation that makes this consensus impossible. The agreements on the interpretation and specification of the principles are a long way from this presumed universal agreement on their validity, making it impossible to have a consensual understanding of their content. Thus, authors like Engelhardt (1996), taking this formality of the principles to the extreme, claim that work in favour of the patient's beneficence lacks concrete content until it materialises based on the interests and preferences of the patient, which points to the idea of the pre-eminence of the principle of autonomy over the others.

However, the emphasis on values such as human dignity that appear in the UNESCO Declaration or in the principles of European bioethics, serve as a common and basic framework for the specific claims of the principles. Promoting the autonomy of patients is not possible without safeguarding their dignity and integrity, and seeking their well-being is not possible without addressing their vulnerability, which is part of the human condition. Acknowledgment of these presuppositions effectively involves emphasising key aspects of an ethics that seeks to be universal. It is necessary to delve into the concepts that support this canon of morality, so that the principles do not remain an empty formality.

Also, we should not lose sight of the constricted nature of the principles. These are conquests of human reason that have been articulating a series of keys that support the suitability of actions and decisions. The principles operate as basic guide that is essential but that is indebted to an historical task of dialogue, proposal, criticism and revision. They cannot therefore be established in a static or immutable manner, as this would lead to paradoxes and contradictions with the passing of time. While it is possible to speak of values that, today, are essential due to having reached a consensus on their importance and irreducibility, and that do hold this title of common morality—such as the dignity of people—the principles are dynamic formulations that are always open to further revision.

In light of the above, the four principles approach, as we have mentioned, is essential. Since they were proposed they have been argued against and worked with, but it is not possible to avoid their influence. Whether it be to analyse the foundations of their validity, or to question their roots in shared values, or to specify and determine their material content for their application in specific situations, or to moderate and contain the casuistic approaches that might lead to arbitrariness, the fact is that the principles of bioethics are a fundamental key to this discipline and, in all likelihood, we will continue debating and reflecting on them in the future.

References cited

Beauchamp, T.L. and J.F. Childress. 2019. Principles of Biomedical Ethics. Oxford University Press, New York.

Beauchamp, T.L. 2020. Principialismo bioético y biojurídico: ¿necesitan la bioética y el bioderecho europeos un marco diferente de principios? Revista Principia Iuris 7: 10–33.

Campbell, A.V. 2000. Uma visão internacional da bioética. pp. 25–35. *In*: Garrafa, V. and S.I. Ferreira Costa (eds.). A bioética no século XXI. Editora UnB, Brasilia, Brazil.

Christen, M., C. Ineichen and C. Tanner. 2014. How "moral" are the Principles of biomedical ethics? A cross-domain evaluation of the common morality hypothesis. BMC Med. Ethics 17: 15–47.

Clouser, K.D. and B. Gert. 1990. A critique of principlism. J. Med. Philos 15: 219–236.

DuBose, E.R., R. Hamel and L.J. O'Connell. 1994. A Matter of Principles? Ferment in U.S. Bioethics. Valley Forge. Trinity Press, Pennsylvania.

Engelhardt, H.T. 1996. The Foundations of Bioethics. Oxford University Press, New York.

Ferrer, J.J. and J.C. Álvarez (eds.). 2003. Para fundamentar la bioética. Teorías y paradigmas teóricos en la bioética contemporánea. UPC-Desclée De Brouwer, Bilbao, Spain.

Ferrer, J.J. 2016. Bioéticas principialistas. pp. 91–116. *In*: Ferrer, J.J., J.A. Lecaros Urzúa and R. Molins Mota (eds.). Bioética: el pluralismo de la fundamentación. Universidad Pontificia Comillas, Madrid.

Frankena, E.K. 1963. Ethics. Englewood Cliffs. Prentice Hall, New Jersey.

Gert, B., C.M. Culver and K.D. Clouser. 1997. Principlism. pp. 71–92. *In*: Gert, B. (ed.). Bioethics: A Return to Fundamentals. Oxford University Press, New York, USA.

Gert, B., C.M. Culver and K.D. Clouser. 2006. Bioethics: A Systematic Approach. Oxford University Press, New York.

Gillon, R. 1994. Medical ethics: four principles plus attention to scope. BMJ 309: 184–188.

Gillon, R. and A. Lloyd. 1994. Principles of Health Care Ethics. John Wiley & sons, London.

Gracia, D. 1989. Fundamentos de bioética. Eudema, Madrid.

Gracia, D. 1991. Procedimientos de decisión en ética clínica. Eudema, Madrid.

Gracia, D. 1997. Cuestión de principios. pp. 19–42. *In*: Feito Grande, L. (ed.). Estudios de bioètica. Dykinson, Madrid, Spain.

Gracia, D. 2011. Teoría y práctica de la deliberación moral. pp. 101–54. *In*: Feito, L., D. Gracia and M. Sánchez (eds.). Bioética: el estado de la cuestión. Triacastela, Madrid, Spain.

Holm, S. 1995. Not just autonomy: the principles of American biomedical ethics. J. Med. Ethics 21: 332–8.

Jonsen, A.R. and S.E Toulmin. 1988. The Abuse of Casuistry. A History of Moral Reasoning. University of California Press, Berkeley.

Jonsen, A.R. 1991. Of balloons and bicycles, or the relationship between ethical theory and practical judgement. Hastings Center Report 21: 14–16.

Mitchell, L.A. 2014. Major Changes in Principles of Biomedical Ethics. A Review of Seven Editions of Beauchamp and Childress. The National Catholic Bioethics Center 14: 459–475.

National Commission for the Protection of Human Subjects of Biomedical and Behavioral Research. 1978. The Belmont Report: Ethical Principles and Guidelines for the Protection of Human Subjects of Research. Bethesda, Maryland.

Neves, M.C.P. 2019. A fundamentação antropológica da bioética. Bioética. 4: 7–16.

Pellegrino, D.E and D.C. Thomasma. 1993. The Virtues in Medical Practice. Oxford University Press, New York.

Pellegrino, D.E. 1994. The metamorphosis of medical ethics. A 30-year retrospective. Arch. Patol. Lab Med. 118: 1065–1069.

Rawls, J. 1971. A Theory of Justice. Harvard University Press, Cambridge.

Rentdorff, J.D. and P. Kemp. 2000. Basic Ethical Principles in European Bioethics and Biolaw. 2 vols. Centre for Ethics and Law & Institut Borja de Bioètica, Copenhagen/Barcelona.

Richardson, H.S. 2000. Specyfing, balancing and interpreting bioethical principles. J. Med. Philos. 25: 285–307.

Ross, W.D. 2002. The Right and the Good. Oxford University Press, Oxford.

Tealdi, J.C. 2005. Los principios de Georgetown: análisis crítico. pp. 36–54. *In*: Garrafa. V., M. Kottow and A. Saada (eds.). Estatuto epistemológico de la bioética. UNESCO, Mexico City, Mexico.

UNESCO. 2006. Declaración Universal sobre bioética y derechos humanos [https://unesdoc.unesco.org/ark:/48223/pf0000146180_spa].

Valdés, E. 2019. Towards a new conception of biolaw. pp. 41–58. *In*: Valdés, E. and J.A. Lecaros (eds.). Biolaw and Policy in the Twenty First Century: Building Answers for New Questions. Springer, Cham, Switzerland.

Applied Bioethics

Chapter 4
Bioethics in Biomedical Research

Magí Farré[1] and *Josep E. Baños*[2,*]

Introduction

Ethical conflicts have always been present in clinical research. However, social awareness of these conflicts arose only after the atrocities that occurred during the Second World War came to light. Chapter 2 of this book outlines the historical path of the relationship between clinical research and bioethics.

Early biomedical research was not considered conflictive. Researchers were convinced that their activity was justified by the significant benefits it would bring to humanity, and participants were satisfied with the compensation they received. However, this scenario started to change when society began to question the model of paternalistic medicine and progressively shifted emphasis toward the rights of individuals. Once the public became aware of what happened during the Second World War, the situation changed completely (Ülman 2009).

Nevertheless, during the second half of the 20th century, bioethics was still not a major concern in biomedical research. Nowadays, it would be inconceivable to undertake research in human beings or even in laboratory animals without considering the ethical issues that might be involved in every step of the investigation. Even outside biomedical settings, ethics is ever-present in all research activities involving human subjects.

[1] School of Medicine, Universitat Autònoma de Barcelona, and Department of Clinical Pharmacology, Hospital Universitari Germans Trias i Pujol.
[2] School of Medicine, Universitat de Vic–Universitat Central de Catalunya.
* Corresponding author: josepeladi.banos@uvic.cat

Clinical research uses data obtained from applying experimental protocols to test the validity of hypotheses in the expectation that the conclusions derived from the study can help improve medical care and public health (Farré and Torrens 2007). Unlike in ordinary clinical practice, where applying the standard of care aims to improve the individual patient's health, in clinical research, applying the study protocol does not always aim to benefit participants; often, the benefits of clinical research will be reaped by unknown patients at an indeterminate time in the future (Beauchamp and Childress 1979). However, it can be difficult to draw a line between ordinary medical practice and clinical research. For instance, some epidemiological studies analyse data obtained from clinical practice, and it is difficult to establish whether studies without interventions on patients should be considered clinical research.

The difference between studies testing an intervention outside established practice and those based on observations with no changes to established practice is crucial. Whereas observational studies require patients to be treated in accordance with clinical practice guidelines or established protocols without added interventions, clinical research applies interventions that are not included in routine care. Some of these interventions (e.g., completing a questionnaire or having more blood extracted than what would be required for routine laboratory analysis to enable genetic studies) involve only minimal inconvenience and risk for patients, but others (e.g., administering a new drug or applying a new surgical procedure) involve risks that can have an enormous impact on patients' outcomes.

In this chapter, we will comment on some historical events that can help us understand the relationship between biomedical research and bioethics. Later, we will consider Beauchamp and Childress's (2019) bioethical principles and research carried out in human beings. Finally, we will discuss ethical issues related to experimental animals.

Historical considerations of ethical issues on medical research

The genesis and the importance of the Nuremberg Code

One of the most important references in the history of bioethics is the atrocities inflicted on prisoners in Nazi concentration camps during Second World War, purportedly for medical research to benefit soldiers. However, the use of "human guinea pigs" predates and extends beyond this period and is more complex than is usually accepted (Jones et al. 2016).

The earliest clinical research where ethical aspects were disregarded took place in the late 19th century, when the aetiology of infectious diseases had been discovered and Western medicine was gaining confidence in applying the scientific method to research in humans (Lederer 1995), although a few prestigious scientists spoke up against unethical practices. In his well-known *Introduction à l'étude de la médecine expérimental,* Claude Bernard (1813–1878) questioned the limits of experimenting on humans: "The principle of medical and surgical morality consists in never performing on man an experiment which might be harmful to him in any extent, even though the result might be highly advantageous to science" (Bernard 1865).

The birth of bacteriology brought about an understanding of the aetiology of diseases that had plagued humanity since the beginning of time, claiming many lives. But this understanding and the thirst for new knowledge led some scientists to practices that resulted in moral outrage. In 1892, Albert Neisser (1855–1916), a professor of dermatology in Breslau who had discovered *Neisseria gonorrhoeae,* the pathogen responsible for gonorrhoea, caused a scandal when he injected unwitting patients with serum from syphilitic patients with the objective of inducing immunity against the disease. Neisser provided patients with no information about the procedure and did not ask for their consent. When some patients developed syphilis, he argued that they were all prostitutes who had acquired the disease during sexual intercourse. When these facts came to light, Neisser was reprimanded and fined for not obtaining consent.

Although Neisser's case caused a scandal and damaged his professional reputation, it is likely that clinical research was often performed without appropriate attention to ethical issues. At the beginning of the 20th century, William Osler, a British doctor working in the United States, wrote that investigators should include a patient in their research only if "direct benefit is likely" and only after obtaining "full consent". He held that unless these requirements were strictly met, "the sacred cord which binds physician and patient snaps instantly" (Osler 1907). A few years later, Walter Cannon urged the American Medical Association to make informed consent compulsory in research (Cannon 1916). The organization refused to act, arguing that bad behaviour was a problem of rogue physicians rather than of research itself. Trust rather than regulation was considered the key issue in improving research and medical care. Attitudes and practices in Europe were similar. In his book *Etyka Lekarska* (Medical Ethics) published in 1917, Teodor Heiman (1848–1917) severely criticized the use of human beings as experimental subjects, especially without seeking their consent to participate in research. He also noted that some desperate patients participated in studies out of fear of their physicians withdrawing their care if they refused to take part (Moreno et al. 2017).

In Germany, concerned that practices such as those that resulted in the scandal surrounding Neisser's case were not exceptional, the

Prussian Ministry for Religious, Educational and Medical Affairs issued an edict to the directors of hospitals and clinics stating that experiments in human beings should be done only after providing them with adequate information about the research and obtaining their consent (with the exception of children and others who could not adequately process the information to provide consent). Furthermore, all research procedures were to be carried out under the direct supervision of the directors of health centres. Promulgating this edict is considered the first attempt to regulate ethical principles in research with human beings.

Initially, the Ministry's edict excluded diagnostic procedures and vaccines. However, in 1931, after 76 neonates in Lübeck died after being injected with a new BCG vaccine against tuberculosis that was probably contaminated with Koch bacilli, the German Ministry issued a new order that made it compulsory to provide information to and obtain consent from people invited to participate in any investigation, regardless of whether its purposes were therapeutic or other.

Unfortunately, however, Prussian laws did not avoid the use of inappropriate practices during the Third Reich. Thousands of prisoners from concentration camps were killed in experiments of supposed military interest. The Nuremberg Doctors' Trial, held between the 9 December 1946 and 19 July 1947, judged 23 people for these actions: 17 were found guilty, and 7 of these were sentenced to death and 9 to long prison sentences (Annas and Groodin 1992).

Less well known, but no less abominable, were the experiments done in Manchuria (China) by physicians in association with the Japanese military during the Second World War. Unit 731 carried out multiple experiments testing agents for chemical and bacteriological warfare on prisoners of war and on the local civilian population, killing at least 12,000 people, mostly Chinese and Russians. These experiments continued until the end of the war stopped them in 1945. Unlike those involved in experiments in Nazi concentration camps, most members of Unit 731 were allowed to continue exercising their professions after the war, probably because agreements between the United States and Japanese authorities sought to procure knowledge from the experiments for allied biological weapons programmes that were considered strategic during the Cold War (Harris 2002, Dhai 2014).

One of the most important consequences of the Nuremberg trials was the publication of the Nuremberg Code in 1947, which aimed to establish what research practices were permissible with human beings. In fact, as Andrew Ivy, one of the main witnesses for the prosecution admitted in his testimony, no formal declaration of ethical principles in clinical research existed before the trial (Shuster 1997). Andrew Ivy and Leo Alexander prepared a ten-point memorandum, "Permissible Medical

Experimentation", that later became the Nuremberg Code (Dhai 2014). Regulations of human research resulted from the acknowledgement that the widely accepted principles expounded in the Hippocratic Oath and the principle of *Primum non nocere* needed to be expanded and brought up to date (Shuster 1997). The Nuremberg Code included broad, absolute statements of the requirement of informed consent (Principle 1) and participants' right to withdraw from the study at any time (Principle 9). The Code needed to go beyond the Hippocratic Oath, which considered medical care, but not medical research (see Chapter 2). Indeed, in research, adhering to the Hippocratic Oath might have undesirable effects, as the physician-defined principle of beneficence could contravene patients' autonomy, as had occurred in patient-doctor relationships for many centuries (Shuster 1997). In this direction, one important contribution of the Nuremberg Code was balancing the need to protect patients stipulated in the Hippocratic Oath (principles 2–8, and 10) with patients' right to protect themselves through their choices (principles 1 and 9) (Shuster 1997).

The Nuremberg Code did not eliminate unethical clinical research, and several studies violating its principles were carried out in the years following its issuance. Perhaps the best known of these was the so-called Tuskegee experiment (Tuskegee Syphilis Study) in the United States, which started in 1932, before the Code was elaborated, but continued long afterward (Reberby 2009). With the aim of delineating the natural course of syphilis, this study enrolled nearly 400 African American patients with syphilis without informing them that they were participating in a research study while withholding penicillin, a drug whose curative efficacy for syphilis was confirmed in the late 1940s. The study remained hidden from the public until it was leaked to the press in November 1972, resulting in a public scandal that led to its termination.

A similar scandal erupted in 2010 when it was revealed that the US Public Health Service, in collaboration with the Guatemalan Health Agency, had financed a study in Guatemala in the 1940s in which soldiers, prisoners, and patients in psychiatric hospitals were infected with syphilis and gonorrhoea (Lerner and Caplan 2016). A series of hearings to investigate these scandals and other unethical experiments (e.g., using mustard gas on soldiers who were pressured to participate without being informed of the risks during the Second World War) (Smith 2008, Jones et al. 2016) resulted in the United States Congress creating the National Commission for the Protection of Human Subjects in Biomedical and Behavioral Research to oversee and regulate medical research in humans. This commission would go on to issue the Belmont Report (see below).

The Declaration of Helsinki: beyond the Nuremberg Code

The Nuremberg Code went on to influence various documents, such as the Universal Declaration of Human Rights adopted by United Nations Organizations in 1948. Importantly, however, the Code never gained full acceptance in some research settings, where it was considered that its origin in reaction to anomalous experiments in the Nazi totalitarian regime that could not be compared with activities in democratic countries made it irrelevant (Dhai 2014). Consequently, many countries continued to allow human research without taking the Nuremberg Code into account. This situation continued until the World Medical Association (WMA) considered it imperative to act, issuing an updated version of the Hippocratic Oath, the Declaration of Geneva in 1948 and the International Code of Medical Ethics in 1949. These documents paved the way for the Declaration of Helsinki, which the WMA General Assembly approved in 1964 and amended in 1975 in Tokyo, in 1983 in Venice, in 1989 in Hong Kong, in 1996 in Somerset West, in 2000 in Edinburgh, in 2008 in Seoul, and in 2013 in Fortaleza (Dhai 2014).

The original Declaration of Helsinki had three sections and 32 paragraphs devoted to different subjects (Fischer 2006). Section A was devoted to defining human research and underlining the need for physicians to prioritize participants' health. Section B commented on the basic principles of clinical research, emphasizing one of the basic tenets of the Nuremberg Code, the need for the study to be grounded in the best available scientific evidence. Finally, section C discussed research that takes place in the context of medical care, stating that research in this situation is acceptable only when the study serves to provide prophylaxis, diagnosis, or treatment. Section C contains two of the most disputed paragraphs: 29, which advises against using placebos as control treatment, only accepting their use when the risk is minimal, the patient's clinical condition is not severe, or when placebos are considered scientifically necessary; and 30, which states that all patients should benefit from the best treatment available once the study concludes.

The Declaration of Helsinki is now internationally considered the best ethical code for clinical research, and its principles are generally accepted and included in national laws governing clinical research.

When declarations are not enough: the works of Pappworth and Beecher

Nevertheless, many clinical investigators in many countries continued to resist conforming to the ethical principles of the Declaration of Helsinki (Jones et al. 2016). This situation only really started to change after new scandals came to light, decreasing public confidence in physician-led clinical research.

The first of these important scandals was the revelation in 1964 that investigators from Memorial Sloan Kettering Cancer Center had injected 22 patients at the Jewish Chronic Disease Hospital in Brooklyn, New York, with cancer cells without asking for their permission. Despite the media scandal, none of the investigators was severely sanctioned for their involvement in this study (Lerner 2004). More instrumental in advancing the process of change were the efforts of two investigators, Maurice Pappworth (1910–1994), an Englishman, and Henry Beecher (1904–1976), an American, who called attention to a series of clinical experiments carried out in their respective countries without informed consent that were published in top scientific journals in the late 1950s and early 1960s. Pappworth and Beecher showed that, even if ethics committees use ethical codes for decision making, these codes are ineffective and insufficient unless they are backed up by specific laws and enforced by institutional control mechanisms. As a consequence of their work, the Department of Health in the United Kingdom established the Local Research Ethics Committees in 1984 based on the principles of the Declaration of Helsinki.

Perhaps the publication that had the greatest impact was Beecher's 1966 paper "Ethics and Clinical Research" reporting on 22 published clinical studies that did not employ ethically acceptable approaches, including one where the hepatitis virus was injected into mentally retarded children to determine the period of infectivity, another in which a mother's melanoma cells where injected into her daughter, and still another where nitrogenous substances were administered to cirrhotic patients to study their relationship with hepatic encephalopathy (Beecher 1966).

Beecher's paper fundamentally changed the way clinical research was performed after the Second World War and also had an impact on the genesis of the Belmont Report. However, as Stark (2016) pointed out, Beecher was not calling for external regulation of the medical profession; rather, he advocated increased self-regulation that would enable physicians to become better investigators through ethically appropriate behaviour. He sought to reform the medical profession from the inside out, not more external regulations. Nonetheless, the most important outcome was contrary to this aim, as his work showed policymakers and the public that physicians did not deserve their autonomy in research and that external regulation was clearly needed. Beecher's paper also prompted some members of Congress to write to the directors of the National Institutes of Health asking for information about the corrective measures to be taken, thus laying the groundwork for the Belmont Report (Rothman 1991).

Another of Beecher's important contributions was his insistence that editors of scientific journals reject manuscripts of studies that did not fulfil the appropriate ethical requirements. The International Committee of Medical Journal Editors, better known as the Vancouver Group, included this point in the Uniform Requirements for Manuscripts Submitted to Biomedical Journals, first published in 1978 and reviewed several times in following years. Among other ethical requirements, this document covers conflicts of interests, the need to maintain participants' privacy and confidentiality, and the need for authors to explicitly state that the research was carried out following local ethical standards and the principles of the Declaration of Helsinki.

The Belmont Report

Pappworth's and Beecher's denunciations did not put an end to unethical clinical research. The above-mentioned Tuskegee study was one of the most important cases. When the news of this study broke, Senator Edward Kennedy led a series of hearings in a congressional investigation into human research that led to the National Research Act in 1974 and the United States Congress creating the National Commission for the Protection of Human Subjects in Biomedical and Behavioral Research (Jones et al. 2016). This commission comprised members trained in ethics, theology, moral philosophy, and religious ethics, and its main objective was "to identify the basic ethical principles that should underlie the conduct of biomedical and behavioral research involving human subjects and to develop guidelines which should be followed to assure that such research is conducted in accordance with those principles" (Jonsen 2004:38). Released in 1978, the Belmont Report was the commission's most important achievement. Not only did the report underline the importance of the well-known principles of respect for persons, beneficence, and justice (Chadwick and Wilson 2018), it also converted these principles into practical operations such as informed consent, risk-benefit analysis, and measures to ensure fairness in selecting participants for clinical research.

From vague indications to concrete specifications: the long road toward regulating bioethics

Over time, the different declarations influenced how bioethical principles were considered in medical research. The Nuremberg Code established the importance of obtaining participants' voluntary consent after fully informing them about the research, of justifying the appropriateness of investigations given what is known, and of ensuring that the investigators are competent to perform the study. The Declaration of Helsinki gained international recognition and went on to become the fundamental code regulating clinical research. The Belmont Report not only laid out the basic bioethical principles, but also specified how to ensure compliance through procedures and documents. The 1975 update of the Declaration of Helsinki stipulated that studies should be based on a written protocol including explicit information about these principles and that this protocol should be reviewed by an independent committee whose members are not involved in the research project (Farré and Torrens 2007).

In 1982, the World Health Organization (WHO)–Council for International Organizations of Medical Sciences (CIOMS) published international guidelines for biomedical research, which were updated in 2016 (CIOMS 2016). In 1995, representatives from regulatory agencies of the European Union, United States, Japan, and other countries met at the International Conference on Harmonisation (ICH) of Technical Requirements for the Registration of Pharmaceuticals for Human Use, elaborating the international harmonised tripartite Guideline for Good Clinical Practice [CPMP/ICH/135/95] (EMA 1996). This important document defined the functions of ethics committees in evaluating clinical trials. The following year, the WHO published the Operational Guidelines for Ethics Committees that Evaluate Biomedical Research (WHO 2011). That same year, under the initiative of the Council of Europe, 40 countries agreed to approve the Convention for the Protection of Human Rights and Dignity of the Human Being with regard to the Application of Biology and Medicine, also known as the European Convention on Bioethics or the Oviedo Convention. Also in 1997, the United Nations Educational, Scientific and Cultural Organization (UNESCO) adopted the Universal Declaration on the Human Genome and Human Rights (UNESCO 1997). In 2005, UNESCO adopted the Universal Declaration on Bioethics and Human Rights (UNESCO 2005) and published Guide number 1 Establishing Bioethics Committees (UNESCO 2005).

Practical considerations of bioethical principles in clinical research

Tom Beauchamp and James Childress elaborated on the principles developed in the Belmont Report in their 1979 book "The Principles of Biomedical Ethics", laying out a clear and simple procedure to determine compliance with bioethical principles in clinical research. Moving beyond the confusing debates of traditional philosophical approaches to bioethics, they proposed a new ethical logic, principlism, that was based on four bioethical principles: autonomy/respect for autonomy, beneficence, non-maleficence, and justice. Principlism gained acceptance among physicians and biomedical scientists, and it has become the most widespread approach to evaluating human research in the United States and in many other countries (see Chapter 3).

The four principles proposed by Beauchamp and Childress (1979) are not all equally important. Justice and non-maleficence are considered the preeminent priorities that must met in all research protocols; by contrast, autonomy and beneficence are considered lesser priorities that might in some circumstances be superseded by the need to fulfil one of the high-priority principles (Beauchamp and Childress 1979, Gracia 1999).

The principle of justice encompasses fairness, compelling researchers to treat all human beings with the same respect. In the selection of participants for a research project, this principle requires researchers to avoid discrimination for any reason and to protect individuals belonging to vulnerable groups such as racial minorities, poor people, women, people with low levels of schooling, or institutionalized people. This principle also requires compensations to be established in case participants are injured. It is important to select the most appropriate participants for the aims of the investigation and for the application of its results.

The principle of non-maleficence consists of refraining from harming others. Researchers must strive to ensure that participants are not harmed in any way. It is embodied in the Hippocratic dictum *primum non nocere* (first do no harm) long used in the practice of medicine. In practical terms, this principle means that investigators have the moral obligation to avoid any injury that might be expected. This is an absolute, wide-ranging principle. For example, including human beings in a study that lacks scientific validity can be considered maleficent. To be considered scientifically valid, a study must be scientifically sound, based on a plausible hypothesis, employing appropriate methods in its design and execution, and enrolling a sufficiently large sample to enable valid conclusions.

The principle of autonomy refers to treating people like independent beings and to respecting their opinions and choices. At the practical level, this means that their consent should be obtained before including them in a scientific study. Participants' consent is valid only if it is given after receiving adequate, understandable information, is given voluntarily, and can be revoked at any time without penalties. The concept of autonomy is closely related to consent and capacity. Undue influence on participants' decision-making processes must be avoided to ensure their decisions are made freely without manipulation, coercion, or deceit. People who are not autonomous or mentally competent (e.g., minors, comatose people, and some people with mental disorders) need special protection, and consent must be obtained from their legal guardians or representatives. The principle of autonomy also requires respecting the right to privacy and the confidentiality of personal data in accordance with each country's national laws. In the European Union, these aspects are regulated by Regulation (EU) 2016/679 of the European Parliament and of the Council of 27 April 2016 on the protection of natural persons with regard to the processing of personal data and on the free movement of such data.

Like the principle of non-maleficence, the principle of beneficence requires researchers to protect participants from harm and to minimize their exposure to risk of injury, but unlike non-maleficence, beneficence also requires that the expected benefits of research be maximized. Thus, beneficence requires a good scientific foundation before undertaking new research, including the quantification of potential risks derived from previous studies in animals and human beings. Studies in which participants cannot expect to gain therapeutic benefits (e.g., pharmacological studies in healthy volunteers) are special cases requiring extreme caution to minimize the possibility of risks or inconvenience. These cases require very strict inclusion and exclusion criteria, continuous monitoring of participants to enable quick action if problems arise, and long-term follow-up if necessary (Beauchamp and Childress 1979, Gracia 1991, Galende 1993).

The concept of informed consent

Clinical study protocols should always include a document with the information that will be provided to participants. This document must include information about the study's objectives, its methodology, the potential risks and benefits involved in participating, how participants' confidential personal data will be safeguarded, participants' right to withdraw their consent and drop out of the study, a statement that choosing not to participate will not affect patients' relationships with their physicians or medical care, an explanation of the insurance policy that will cover the treatment of eventual injuries, and a declaration of participants' right to ask investigators for additional information (Farré and Torrens 2007). Table 1 summarizes the Spanish Medicines Agency's recommendations for the minimal contents to be included in the information sheet provided to patients invited to participate in a clinical trial. Furthermore, the information must be conveyed in language that can be easily understood by participants who have completed elementary school. All necessary information must be included, and the risks and benefits of participation must be fully discussed. However, it is equally important not to include unnecessary information that might confuse rather than inform participants.

It is essential to ensure that possible participants have understood all the points in the information sheet. Investigators must confirm that the participant understands what being included in the study entails. Potential participants should be afforded sufficient time to think about the study and to query investigators and consult others if they consider it helpful before giving their final decision. Ideally, potential participants will receive the information sheet at least one day before delivering their decision about whether to consent.

Table 1. Minimal content to be included in the information sheet provided to participants in human research projects or clinical trials (Spanish Medicines Agency. Guideline for correct preparation of a model patient information sheet and informed consent form, available at: https://docs.google.com/viewer?url=https%3A%2F%2Fwww.aemps.gob.es%2FinvestigacionClinica%2Fmedicamentos%2Fdocs%2Fannex8a-Ins-AEMPS-EC.pdf).

- Study title, study code, EudraCT number, sponsor, principal investigator, site, version, and date of the research project
- Introduction
- Voluntary participation
- Study objective
- Study description
- Study activities
- Risks and inconveniences of participating in the study
- Possible benefits
- Contact in case of questions
- Pregnancy warning
- Alternative treatments
- Expenses and financial compensation
- Treatment to be received when the clinical trial ends
- Insurance
- Personal data protection
- Explanation of how data will be used
- Other relevant information
- Clinical studies in minors (when applicable)
- Collection and use of biological samples
- Substudies
- Informed consent/Witnessed informed consent

When there are reasonable doubts about the ability of a potential participant to understand the information about the study (e.g., participants with decreased mental capacities), it is compulsory to involve their legal representatives. Potential participants who are illiterate or have difficulties understanding written information should receive all the information orally in the presence of a witness. If, despite these measures, the investigator remains uncertain whether a potential participant has fully understood the information, it is advisable to decline their participation.

It is also essential to ensure that participation is voluntary. The physical presence of the investigators, especially if they are involved in treating the potential participant, may have a coercive effect. Leaving adequate time for reflection or consultation with people they trust can help potential participants feel confident in their decision.

Human research ethics committees

Research protocols need to be independently evaluated from the scientific and ethical points of view. The need to ensure that clinical research protocols are based on sound science and comply with ethical requirements led to the creation of ethics committees for clinical research in many countries. Ethics committees are multidisciplinary, independent groups of individuals appointed to review biomedical research protocols involving human beings to help ensure that the dignity, fundamental rights, safety, and well-being of research participants are duly respected and protected. These committees may be established at the local, regional, or national level (Council of Europe 2012).

Ethics committees include members of different professions, such as physicians, other health professionals, lawyers, and laypersons from outside the institution; if the characteristics of the study warrant it, additional health or social professionals can be invited to join the committee. Investigators cannot participate in evaluating the protocols of studies they are involved in or try to influence the committee's final decision. Members who are involved in a study in any way or have other conflicts of interest must recuse themselves from evaluations, and their recusal must be noted in the minutes of the meeting (Council of Europe 2012, World Health Organization 2011).

The committees must evaluate ethical and methodological aspects of the research protocols together, as these two aspects are inseparable. A methodologically sound study may or may not be ethically correct, but a methodologically unsound study can never be ethically correct. Methodological aspects are usually evaluated by members who are health professionals, and ethical aspects are usually evaluated by members who are not health professionals.

Genetic research

It is becoming increasingly common to obtain biological samples in all human studies to carry out genetic analyses to determine the influence of genes and their mutations on pathogenesis or on the therapeutic or toxic effects of drugs. All tissue samples, cells, or isolated DNA contain the participant's genetic code and therefore require special treatment. Genetic material contains private information about an individual, and because this information is transmissible and human cloning is possible, we can say that genetic material even transcends the individual (UNESCO 2003).

Genetic research requires specific legislation to protect participants from the inappropriate use of their biological material. Some of these aspects are considered in UNESCO's Universal Declaration on the Human Genome and Human rights (UNESCO 1997) and in the Council of Europe's

1997 Convention for the Protection of Human Rights and Dignity of the Human Being with regard to the Application of Biology and Medicine, which forbids interventions aiming to create a human being identical to another and thereby human cloning. Thus, it is compulsory to inform subjects and obtain their consent whenever genetic studies might be done on the samples they provide. Table 2 lists the minimum elements that this consent must include.

Table 2. Minimal content to be included in information sheets for informed consent in genetic studies (Law 14/2007, of July 3rd, on Biomedical Research, Spain. Available at: https://www.boe.es/buscar/doc.php?id=BOE-A-2007-12945).

- Purpose of the genetic analysis, including the genes that will be studied
- Place where the analyses will be done and destination of the biological sample afterwards
- Persons who will have access to the results of the analysis when they not be disassociated or anonymised from the participant
- A warning about the possibility of unexpected findings and their possible transcendence for the participant, as well as their right to decide whether to receive this communication
- A warning about the implications that the information obtained might have for the participant's family members and that the participant, where appropriate, will have to decide whether to convey that information to them
- Commitment to providing genetic counselling once the results of the analysis are evaluated
- A statement that the material will be used only in the current study and only for the stated purposes, adding that additional consent will be sought if researchers wish to use this material in other studies or for other purposes
- Details about how samples will be treated, how confidentiality will be guaranteed, how long samples will be kept (maximum time), and whether samples might be transferred to a third parties
- Whether the research might result in the obtainment of patents

Animal research and ethics

Animal experimentation also requires adherence to bioethical principles. Although this chapter concentrates on ethics in clinical research, some comments on how to apply ethical principles to research with animals are also warranted. Centres that experiment on animals must have specific ethics committees to oversee their work. Analogous to ethics committees for human research, animal research ethics committees aim to guarantee responsible research with animals other than humans, advising researchers on animal welfare issues (e.g., acquisition, housing, care, and use).

In 1959 William Russell and Rex Burch published The Principles of Humane Experimental Technique, the first (and for many years, the only) textbook on how to apply ethical principles in animal research (Russell and Burch 1959). This seminal treatise which established the "three Rs" principle, named for three actions aiming to minimize animal experimentation and distress during essential experiments: replacing, reducing and refining.

The first principle, replace, calls for using approaches that do not require animals whenever possible, and using the least sentient animals possible when experiments using animals cannot be avoided. The second, reduction, calls for using the smallest number of animals needed to enable statistical analyses. And the third, refining, calls for using the best techniques to reduce animal suffering, not only with regard to experimental techniques but also to housing and other measures to minimize pain, suffering, and distress.

Although applying the three Rs has vastly improved animal welfare, this approach fails to consider other moral principles that are involved in core issues in animal experimentation today (Beauchamp and DeGrazia 2020). Various developments have given rise to new challenges in the bioethics of animal research. Our knowledge and understanding of animal physiology and behaviour have advanced, and the public's perception of animal welfare has also evolved. In the late 20th century, the ethics of animal research became a new area of scientific interest, leading to important changes in the way we think about the moral status of the animals and our responsibility in caring for them and using them in research (Beauchamp and DeGrazia 2020).

In recent years, some authors have maintained that sentient animals have moral rights that preclude them from being considered only as a research interest and that the only possible justification for infringing on these rights or injuring these animals is the obtainment of great social benefits that could not otherwise be achieved. From these positions, two core values are derived: social benefit and animal welfare.

Beauchamp and DeGrazia (2020) advocated for breaking these two core values down into six principles. The first core value, social benefit, involves ensuring that there is no alternative method, analysing the likelihood of expected net benefit, and determining whether the research can yield sufficient value to justify harm. The second core value, animal welfare, involves avoiding unnecessary harm, ensuring that animals' basic needs are met, and establishing the upper limits of harm that can be inflicted.

The application of these six principles has facilitated significant progression toward research that is more respectful of experimental animals' well-being, although difficulties in applying them remain.

What does the future hold for bioethical principles in research?

Fifty years after Beecher's pivotal article, Jones et al. (2016) reflected on the state of bioethics in clinical research, concluding that, despite the formulation of new legal regulations and increased sensitivity and knowledge about what is ethical and what is not, studies violating recognized bioethical principles have probably not been eradicated. These authors relate three lessons learned after 150 years of clinical research. The first is that the evidence shows that ethical values change over time, and it is important to understand why and how. The second is that it is difficult to reach a full consensus about what ethics is and what it is not, even among professionals who are versed in the issue. And the third is that many interests could tempt investigators to carry out studies even when they aware that they are inappropriate. For these reasons, the chances that unethical studies will be performed in the future are not negligible, and tireless supervision is required to prevent transgressions of ethical principles in research.

One important issue, mentioned in the updates to Declaration of Helsinki, is to increase transparency in clinical research. Measures to make research more transparent include registering studies in publicly accessible databases (e.g., clinicaltrials.gov) before recruitment starts, paying attention to ethical obligations in the publication and dissemination of research results (e.g., by including negative and inconclusive as well as positive results and making data available to other researchers), and declaring sources of funding, institutional affiliations, and conflicts of interest.

In 2000, Emanuel et al. (2000) published an article entitled "What Makes Clinical Research Ethical?" Their answer to the question posed in the title included some of the aspects discussed above, as well as additional elements, summarised and explained in the following seven points:

1. *Social or scientific value*. Studies must improve health and well-being or increase knowledge. Without social or scientific value, research exposes participants to risks for no good reason and wastes resources.
2. *Scientific validity*. Research must be conducted in a methodologically rigorous manner that is feasible and produces reliable and valid data that can be interpreted.
3. *Fair subject selection*. Selection of subjects must not be discriminative and must protect vulnerable individuals.
4. *Favourable risk-benefit ratio*. Researchers must minimize risk and maximize benefits. Clinical research must be conducted in a manner consistent with the standards of clinical practice.
5. *Independent review*. Research protocols must be evaluated by an independent ethics committee.

6. *Informed consent.* Subjects must be adequately informed and provide their consent voluntarily.
7. *Respect for potential and enrolled subjects.* Researchers have ongoing obligations to participants (e.g., permitting withdrawal from the study and protecting privacy through confidentiality).

Some years later Emanuel added another principle for investigation in developing countries, although it can also be applied to research in other contexts (Emanuel 2004):

8. *Collaborative partnership.* A collaborative partnership between researchers and sponsors, policy makers, and communities could minimize the possibility of exploitation and make research more acceptable for the community.

These eight points provide a framework to guide researchers and members of ethics committees in assessing clinical research protocols. Finally, in recent years concern has arisen that clinical trials might be performed in developing countries to elude control over violations of bioethical principles that are widely accepted in Western countries. Three aspects of research in developing countries warrant special attention (Emanuel et al. 2004, CIOMS 2021): the standard of care in developing countries, the reasonable availability of interventions that are useful in clinical trials, and the quality of informed consent. Participants must be protected from exploitation.

The risk of such exploitation is not merely theoretical. In 2001, the British writer John Le Carré published *The Constant Gardener*, a novel inspired in an illegal clinical trial carried out in children with bacterial meningitis in Nigeria in which some participants died (Raufu 2003, Stephens 2006). To reduce the possibility of situations like this occurring again, the CIOMS recently prepared a document detailing safeguards to prevent such situations (CIOMS 2021). In the 21st century, we must remain vigilant to ensure that unethical experimentation in humans does not occur anywhere.

References cited

Annas, G.J. and M.A. Grodin. 1992. The Nazi Doctors and the Nuremberg Code: Human Rights in Human Experimentation. Oxford University Press, New York.

Beauchamp, T.L. and J.F. Childress. 1979. Principles of Biomedical Ethics. Oxford University Press, New York.

Beauchamp, T.L. and D. DeGrazia. 2020. Principles of Animal Research Ethics. Oxford University Press, Oxford.

Beecher, H.K. 1966. Ethics and clinical research. New Eng. J. Med. 274(24): 1354–1360.

Bernard, C. 1865. Introduction à l'étude de la médecine experimental. Consulted: Bernard, C. 1957. An Introduction to the Study of Experimental Medicine. Dover, New York.

Chadwick, R. and D. Wilson. 2018. The emergence and development of bioethics in the UK. Medical Law Review 26(2): 183–201.
Cannon, W.B. 1916. The right and wrong of making experiments on human beings. JAMA 67: 1372–1373.
CIOMS. 2016. International Ethical Guidelines for Health-related Research Involving Human. Prepared by the Council for International Organizations of Medical Sciences (CIOMS) in Collaboration with the World Health Organization (WHO). CIOMS, Geneva.
CIOMS. 2021. Clinical Research in Resource-limited Settings. A Consensus by CIOMS Working Group. CIOMS, Geneva.
Council of Europe. 1997. Convention for the Protection of Human Rights and Dignity of the Human Being with regard to the Application of Biology and Medicine: Convention on Human Rights and Biomedicine (Available at: https://rm.coe.int/168007cf98).
Council of Europe. 2012. Guide for Research Ethics. Committee Members. Steering Committee on Bioethics Council of Europe. Available in: https://rm.coe.int/CoERMPublicCommonSearchServices/DisplayDCTMContent?documentId=0900001680307e6c.
Dhai, A. 2014. The research ethics evolution: From Nuremberg to Helsinki. S Afr. Med. J. 104(3): 178–180.
Emanuel, E.J., D. Wendler and C. Grady. 2000. What makes clinical research ethical? JAMA 283: 2701–2711.
Emanuel, E.J., D. Wendler, J. Killen and C. Grady. 2004. What makes clinical research in developing countries ethical? The benchmarks of ethical research. J. Infect. Dis. 189: 930–937.
EMA. 1996. European Medicines Agency. Note for guidance on Good Clinical Practice. (CPMP/-ICH/135/95). Available at www.emea.europe.eu/pdfs/human/ich/013595en.pdf.
Farré, M. and M. Torrens. 2007. Aspectos éticos y legales. Comités Éticos de Investigación Clínica. Consentimiento informado. pp. 137–156. *In*: Ballesteros, J., M. Torrens and J.C. Valderrama (eds.). Manual introductorio a la investigación en drogodependencias. Sociedad Española de Toxicomanías, Valencia, Spain.
Fischer, B.A. 2006. A summary of important documents in the field of research ethics. Schizophrenia Bulletin 32(1): 69–80.
Galende, I. 1993. Problemas éticos de la utilización de seres humanos en investigación biomédica. Consecuencias sobre la prescripción. Med Clin (Barc). 101: 20–23.
Gracia, D. 1991. Procedimientos de decisión en ética clínica. Eudema, Madrid.
Harris, S.H. 2002. Factories of Death: Japanese Biological Warfare, 1932–1945, and the American Cover-up. Psychology Press, New York.
Jones, D.S., C. Gradt and S.E. Lederer 2016. "Ethics and clinical research"—The 50th anniversary of Beecher's bombshell. New Eng. J. Med. 374(24): 2393–2398.
Jonsen, A.R. 2004. The history of bioethics as a discipline. *In*: Khushf, G. (ed.). Handbook of Bioethics: Taking Stock of the Field from a Philosophical Perspective. Kluwer Academic Publishers, New York.
Lederer, S.E. 1995. Subjected to Science: Human Experimentation in America before the Second World War. Johns Hopkins University Press, Baltimore.
Lerner, B.H. 2004. Sins of omission—cancer research without informed consent. N Eng. J. Med. 351: 628–630.
Lerner, B.H. and A.L. Caplan. 2016. Judging the past: How history should inform bioethics. Ann. Intern. Med. 164(8): 553–557.
Moreno, J.D., U. Schmidt and S. Joffe. 2017. The Nuremberg Code 70 years later. JAMA 318: 795–796.
Osler, W. 1907. The evolution of the idea of experiment in medicine. Trans. Cong. Am. Phys. Surg. 7: 7–8.

Raufu, A. 2003. Nigerians in drug trial take their case to US court. BMJ 326(7395): 899.
Reberby, S.M. 2009. Examining Tuskegee: the Infamous Syphilis Study and its Legacy. University of North Carolina Press, Chapel Hill.
Rothman, D.J. 1991. Strangers at the Bedside: A History of How Law and Bioethics Transformed Medical Decision Making. Basic Books, New York.
Russell, W.M.S. and R.L. Burch. 1959. The Principles of Humane Experimental Technique. Methuen, London, England.
Shuster, E. 1997. Fifty years later: the significance of Nuremberg Code. New Eng. J. Med. 337: 1436–1440.
Smith, S.L. 2008. Mustard gas and American race-based human experimentation in World War II. J. Law Med. Ethics 36: 517–521.
Stephens, J. 2006. Panel Faults Pfizer in '96 Clinical Trial in Nigeria'. The Washington Post May 7, 2006.
Stark, L. 2016. The unintended ethics of Henry K. Beecher. Lancet. 387: 2374–2375.
Ülman, Y.I. 2009. Coursebook of Medical Ethics and Medical History. Istanbul University, Istanbul.
UNESCO. 1997. Universal Declaration on Human Genome and Human Rights (Available at: https://www.unesco.org/en/legal-affairs/universal-declaration-human-genome-and-human-rights).
UNESCO. 2003. Declaration on Human Genetic Data (Available at: https://www.unesco.org/en/legal-affairs/international-declaration-human-genetic-data?hub=66535).
UNESCO. 2005. Universal Declaration on Bioethics and Human Rights (Available at: https://www.unesco.org/en/ethics-science-technology/bioethics-and-human-rights).
UNESCO. 2005. Establishing Bioethical Committees. Guide Number 1 (Available at: https://unesdoc.unesco.org/ark:/48223/pf0000139309).
World Health Organization. 2011. Standards and operational guidance for ethics review of health-related research with human participants. Guidance documents. Available in: https://www.who.int/publications/i/item/9789241502948.

Chapter 5

Unresolved Problems in Bioethics

The Beginning and End of Life

Núria Terribas-Sala

Introduction

Since the creation of the first Centres for the study of bioethics in the United States more than 50 years ago, bioethicists, scientists, theologians, academics, politicians and society have debated questions concerning the value of human life and, in particular, its beginning and end, without there being anything approaching a consensus on these issues.

Discussion of the value of incipient life, and whether it is possible and legitimate to interfere with it, remains a key focus of bioethics. This entails debate around when human life begins, when the embryo should be protected, what other values need to be taken into consideration, which interventions are legitimate with respect to this incipient life where to do so is in conflict with the health of third parties, research needs, or women's freedom to control their own bodies. Achievements in the understanding of human reproduction, contraception, the possibility of medical interventions to prevent disease in the unborn, and the need for ongoing genetic research have meant that bioethics has constantly had to review arguments and reconsider its positions. At the same time, changes in European society, with low birth rates, delayed maternity,

UVic-UCC Chair of Bioethics, Fundació Grífols, Spain.
Email: nuria.terribas@grifols.com

fertility problems and the appearance of new family models, require us to update our ethical evaluation of all these concepts. Despite these, we remain rooted in a debate that is sometimes excessively polarized, making progress difficult, and contributing to inequity and huge variations between countries.

By the same token, the end of life and the issue as to whether individuals—ill or not—have the right to decide when and how it ends, is another of the questions that has generated a lot of bioethical debate over the decades. This intensified when the individual's right to autonomy in the health sphere was granted legal recognition, enabling fully competent individuals to take their own decisions, whether in the current moment, with fully capacity, or in advance. This has led to changes in medical practice at the end of life with the result that the unremitting defense of biological life by doctors, often where this involved exhausting all the means at their disposal for no good reason, has gradually given way to attitudes based on a respect for autonomy, accepting the rejection of treatment, and adapting decisions at the end of life to include criteria of proportionality, utility and justice. Alongside these changes, in recent years a more radical application of this respect for autonomy has been proposed: recognition of the right to decide how and when to die, with the help of professionals with euthanasia or medically assisted suicide.

It is these two major topics in bioethics—both of which are ongoing—that we address in this chapter.

Historical debate about the value of human life and the influence of religion

Christian tradition in general and Catholicism in particular have played a large role in the bioethical debate about the value of human life, sustaining the dogma of the inviolability of human life, which is given by God and which only God can take away. Indeed, it could be argued that the confrontation between religious and secular culture is a particularly salient feature of European and American bioethics. Despite the secularization that is common throughout Europe, Catholic culture continues to exercise a powerful influence on bioethical debate, in particular where it touches upon issues related to the beginning and end of life. Dialogue between the two cultures is difficult in so far as the Church, after many centuries of influence, is extremely reluctant to accept the separation of its power from that of the State and strives to continue to impose the official doctrine of the Church in a society that is culturally, morally and ideologically pluralistic. This is often seen as an attempt to interfere in the lives of ordinary people and, in some countries, has led to growing support for secularization, seeking to remove all religious manifestations from the public sphere (Abel 2012).

In this respect, it is important to review the interpretation given to the Fifth Commandment—thou shalt not kill—in precept 2258 of the Catechism of the Catholic Church, established by the Congregation for the Doctrine of the Faith (1987), which states, "Human life is sacred because from its beginning it involves the creative action of God and it remains forever in a special relationship with the Creator, who is its sole end. God alone is the Lord of life from its beginning until its end: no one can under any circumstance claim for himself the right directly to destroy an innocent human being".

On this basis, despite the progress in our understanding of the biological and genetic processes of human reproduction, the Catholic Church continues to defend the existence of human life as inviolable from the moment of conception—egg fertilization by the sperm—so that even those interventions designed to prevent or avoid the gametes encountering each other, whether through physical procedures or drugs, are to be condemned. This means that the Catholic Church does not accept contraceptive methods nor, of course, abortion, but nor does it endorse assisted reproduction techniques, in so far as these represent intervention in or interference with the natural reproductive process (Congregation for the Doctrine of the Faith 2008). And the Church continues to insist on similar criteria, reflecting a growing distance between it and modern society and people's needs.

With regard to the end of life, there has been some evolution in the doctrine of the Catholic Church with respect to end-of-life care, although its condemnation of euthanasia or assisted suicide remains a matter of dogma (2277 "Whatever its motives and means, direct euthanasia consists in putting an end to the lives of handicapped, sick, or dying persons. It is morally unacceptable. Thus an act or omission which, of itself or by intention, causes death in order to eliminate suffering constitutes a murder gravely contrary to the dignity of the human person and to the respect due to the living God, his Creator") (The Holy See 1992).

Historically, the suffering associated with illness at the end of life was linked to purification of the soul and becoming closer to God, and this meant that such suffering could not be mitigated by drugs or other means. Pain and the other unbearable symptoms that accompanied the process of dying were considered as something good, causing the individual to appear with greater goodness and glory before the Creator in their final moments. As a result, for many decades the use of opioids or other substances to relieve pain was not permitted, and in many health or hospital services managed by religious orders, their use was restricted, inflicting greater suffering on the dying. This may still be the case in some parts of the world where, furthermore, palliative drugs are scarce and access to them is difficult.

At the same time, in the context of progress in medical technology and the possibility of prolonging people's survival using life support measures, the criterion that life was a sacred gift that must be preserved until the last breath using all possible measures, continued to prevail.

However, the current version of the Catechism of the Catholic Church reflects a view that is more reasonable and is more respectful of the individual's wishes, stating in precept 2278: "Discontinuing medical procedures that are burdensome, dangerous, extraordinary, or disproportionate to the expected outcome can be legitimate; it is the refusal therapeutic obstinacy or overtreatment. Here one does not will to cause death; one's inability to impede it is merely accepted. The decisions should be made by the patient if he is competent and able or, if not, by those legally entitled to act for the patient, whose reasonable will and legitimate interests must always be respected" (The Holy See 1992).

And in precept 2279: "Even if death is thought imminent, the ordinary care owed to a sick person cannot be legitimately interrupted. The use of painkillers to alleviate the sufferings of the dying, even at the risk of shortening their days, can be morally in conformity with human dignity if death is not willed as either an end or a means, but only foreseen and tolerated as inevitable. Palliative care is a special form of disinterested charity. As such it should be encouraged" (The Holy See 1992).

Other religions, such as Judaism, Islam and Hinduism, also consider life as a gift that is granted to us and which we must administer correctly but may not dispose of freely. As a result, all actions that enable or assist suicide, whatever the person's situation, are deemed unacceptable (Junta de Andalucía 2008).

All of these concepts were already at the forefront of the first debates between scientists, theologians and jurists at the start of bioethics, and have presented a stumbling block in attempts to reach a consensus based on a civil ethics that puts religious beliefs to one side and enables progress in respecting the individual freedom of those who do not share such beliefs, which should not be imposed on a diverse and morally pluralistic society.

After 50 years of bioethics, this has not been possible, and with respect both to the beginning and the end of life, we continue to find ourselves immersed in debates to which there is no universal solution. Even where common reference frameworks have been established, such as the *European Convention on Human Rights and Biomedicine of the Council of Europe* (Council of Europe 1997) or the UNESCO *Universal Declaration on Bioethics and Human Rights* (UNESCO 2005) it proved impossible to achieve a shared baseline position on these questions. As a result, each country has developed in ways that reflect their specific realities and social demands and are constrained by the ideology of the dominant political forces, often biased by religious belief. In some instances, bioethical debate and a more well-established tradition of deliberative democracy have enabled far-reaching progress and legislative change, such as has been the case in the United Kingdom and, in recent years, in Spain. In other countries, despite the existence of clear social demand (for example, with respect to abortion), political ideology has prevented such change, even though there are often opposing positions within the political parties themselves.

When we look at the question of the beginning of human life, we find a greater degree of openness and more progress, through the regulation in many countries of assisted reproduction or abortion, while when it comes to the end of life, the regulation of euthanasia and medically assisted suicide remains a minority.

The beginning of human life and intervention in it

Positions with respect to the start of human life and the value of the embryo

The bioethical debate about the legitimacy of intervening in incipient human life has focused on an attempt to pin down when human life begins and what protection we should afford this life that is in the process of development. Progress in recent decades in our knowledge of genetics and the importance of epigenetics has been vital in prompting a reconsideration of the value of 'human life' and thus also to reconsider the ethical and legal acceptability of certain actions in the sphere of human assisted reproduction, in its therapeutic and diagnostic possibilities, and in research with embryos or stem cells.

Traditionally, bioethical questions about the value of human life and the ontological status of the human embryo have been (Abel 2012):

- Should the embryo be considered as a person, and from what moment or time of its development?
- When can we talk about the 'dignity' of the human embryo?
- What protection should the law give the embryo?
- What interventions at the beginning of human life are ethically acceptable?

The answers to these questions are conditioned by the position that we adopt with respect to the beginning of human life. We can clearly identify three distinct positions:

a) **Conservative position:** starts from the view that human life begins with the formation of the zygote following fertilization of the oocyte, establishing what will be a new human being with clearly determined biological characteristics. In this view, the moral condition of the human embryo does not depend on an arbitrary moment in its development and the embryo should instead be respected as a 'person' from the first moment of its existence. From this flows the inviolability of this potential person and the right to life of the human being from the moment of conception, rejecting any action that prevents or does not promote this development.

According to this view, which is the one adopted by the hierarchy of the Catholic Church (Congregation for the Doctrine of the Faith 1987), any artificial contraceptive measure (contraceptive pill or other hormonal treatments, intrauterine devices, morning-after pill, etc.) is ethically and morally unacceptable. By the same token, any assisted reproduction technique that interferes with the natural reproductive process is inadmissible, as of course are abortion or any other practice which interrupts pregnancy. Finally, *in vitro* interventions, whether diagnostic or therapeutic, are censured, as is any research activity performed with embryos or stem cells obtained from human embryos.

b) **Intermediate position or gradual protection of the embryo:** considers that the zygote does not yet contain all the elements of the future human being. Embryogenesis, with the embryo's implantation in the mother's uterus, is a period of biological constitution while which new qualities emerge which did not previously exist, either in reality or potentially. The constitutive process is not completed until the embryo has been implanted, and the new being continues to be configured through the interaction between mother and embryo. During this process, which takes days (around 2 weeks), the value of 'human dignity' may be balanced with other, potentially conflicting, values (e.g., the mother's health, malformation of the foetus, etc.). And when this initial process occurs in vitro, values such as the health of the future child, through diagnosis, or the interests of research and science, may be taken into consideration when contemplating the ethical and moral legitimacy of certain actions. As a result, 'human dignity' cannot be ascribed to the in vitro or frozen embryo or to the human embryo until it is fully implanted in the uterus. This does not mean it should have no protection under the law, but such protection should be gradual or proportionate and not absolute, unlike the protection given to the 'human life' of the newborn child.

In this view, interventions designed to prevent fertilization of the oocyte or implantation of the fertilized oocyte, and assisted reproduction techniques designed to promote this fertilization—whether intrauterine (artificial insemination) or *in vitro*—are perfectly legitimate and acceptable.

When it comes to laboratory activity involving embryos that are just a few days old and have been rejected for reproductive use by their progenitors, the interests of the community and of society in improving scientific understanding would be balanced with the value of the frozen embryo with no reproductive viability. It seems morally preferable to give it a beneficial use for society rather than destroying it.

With respect to abortion (the interruption of pregnancy with a foetus that is developing in the uterus), an approach based on gradual protection weighs up the ethical and legal conflict between the right to life of the unborn and other values (mother's health, disability of the unborn child, dignity of the mother in a pregnancy that results from rape, and the woman's freedom to control her own body in the case of abortion on demand) (Institute Borja of Bioethics 2009). In this conflict of values, the law has the task of reaching a decision in one direction or the other, based on respect for a morally plural society (Terribas Sala 2014).

c) **Extreme liberal position:** the view that an embryo in a laboratory is a mere collection of cells with no specific value beyond its possible benefit to society or third parties. In this purely biological view, these embryo cells require no protection beyond that which derives from the regulation of research, which must be performed in accordance with ethical criteria, as in other biomedical spheres. With respect to the intrauterine embryo, the decision about an intervention to interrupt pregnancy must always be taken by the pregnant mother, as the person who is control of her own physical integrity. The most extreme versions of this liberal position argue that there is no 'person' to whom the right to life could be attributed until a being that is fully independent of the mother, and thus no conflict of values between the life of the unborn child and the mother's freedom (Casado et al. 2021).

Interventions involving the human embryo: assisted reproduction techniques, surrogate motherhood, abortion and research

As we have noted, there are many spheres of possible intervention involving the human embryo, all of which are linked to reproduction or research. Bioethical reflection encompasses all of them, and the moral evaluation of different actions may vary depending on the starting point about the value of incipient human life and how this value is balanced with the other values involved.

In this chapter we do not have space to explore these questions in depth, each of which has been debated for many years and has generated its own academic literature. However, we will briefly summarize the different subjects of bioethical debate.

Human assisted reproduction techniques

This heading includes all those techniques designed to help couples or women with fertility problems who want to have children. It covers a range of different methods, depending on the nature of the fertility problem and whether donor genetic material is required. The ethical acceptability of these techniques also varies depending on these parameters. There are some countries which only accept homologous techniques—using gametes from the couple—and reject heterologous techniques—using donor material—on the basis that intervention by a third party breaks the unity of the reproductive project, which should remain within the couple. An ethical distinction is also sometimes drawn between artificial insemination—inside the woman's body—and *in vitro* fertilization, followed by embryo transfer.

This procedure, which has a higher rate of success, requires embryos to be obtained in the laboratory by obtaining eggs, fertilizing them *in vitro*, verifying that fusion has occurred correctly, and then transferring them to the woman's uterus for implantation and development, until birth takes place. This technique typically involves obtaining more than one embryo and freezing any additional embryos for use in future cycles if required. It is a process that involves far more intervention, including the artificial manipulation of embryos, and as a result has met with more resistance in some countries for cultural and religious reasons.

If we add to this the need to use donor gametes as a result of male or female infertility or both, then its ethical acceptability is complicated yet further, according to some concepts of human reproduction.

European legislation on assisted reproduction techniques varies widely, with the United Kingdom and Spain having the most permissive laws. As a result, Spain has one of the highest rates of foreign women visiting the country to use its fertility services, because in their own countries the law is more restrictive.

A second level in the use of assisted reproduction techniques relates to diagnosis and treatment, where in some cases in which there is a family history of hereditary genetic diseases, diagnosis can be performed prior to implantation of embryos in the laboratory to identify whether they carry the genetic disease and to exclude them from use in reproduction. This possibility—for example, in families with hereditary breast cancer (Brca1)—has offered a way of breaking the inter-generational chain of transmission for this kind of cancer. However, not all hereditary genetic diseases can be clearly identified in this way (where the carrier of the mutated gene is certain to develop the disease) and we therefore enter the sphere of statistical probability. Bioethical debate in this area therefore involves asking whether simple prevention on the grounds of statistical probability justifies actions that involve discarding viable embryos.

In this context, a particularly important development from a bioethical perspective is the application of germline genome editing of *in vitro* embryos using the CRISPR9 technique, avoiding discarding the embryo, removing the pathologic gene. This issue is discussed in another chapter of this book.

Surrogate motherhood

The issue of surrogate motherhood is one of the most controversial and has met with broad opposition, although the reality is that it is practiced in some countries.

It is seen as a solution for gay male couples or single men who want to be fathers without having a partner, although it is also resorted to by heterosexual couples in which, for medical reasons, the woman is unable to bear a child. The couple who wish to have a child use intermediate agencies to look for a potential mother and the pregnancy can take a number of forms, depending on the source of the genetic material: from the couple; from one of them, using a donor; or with the oocyte of the surrogate mother (Torres 2019).

Beyond to the technique used or the source of the genetic material, ethical discussion of surrogate motherhood focuses on the use or instrumentalization of the woman as a 'means' to achieve this reproduction, outside of the context of the couple or of the single man who wishes to be a father. We must also mention the aspect of the dignity of the future child, who will be born "on request", of special relevance for the most critical position against surrogacy, when it is the result of an economic transaction. In countries where this is permitted as an act of free choice by the woman (e.g., USA), the debate is focused on monetization, with significant sums of money involved in such contracts, largely to the benefit of the intermediaries, and on the restrictions on the freedoms of the surrogates, who are supervised and constrained throughout the pregnancy. In countries where these practices constitute forms of slavery and exploitation of women (e.g., India), without any freedom to choose and held in virtual captivity during the nine months of the pregnancy under severe conditions, the issue becomes one of blatant violation of human rights (Profesionales por la Ética 2015).

In the United Kingdom, surrogate motherhood has been regulated on the basis that there is no exchange of money between the couple and the pregnant woman, the aim being to guarantee that the process is motivated by solidarity and altruism, without any financial motivation to influence the commitment to become pregnant on behalf of another person, and with maximum respect for the wishes of the pregnant woman throughout the process, including the right to refuse to give up the child after it is born. This method is rarely used as it does not give couples any guarantee that they will achieve their objective—to have a child—except in cases where they are related or have close ties to the pregnant woman (Albert 2017).

However, in a rapidly changing society with a diversity of family models, countries are going to have to regulate this issue, legalizing births by surrogate motherhood arranged abroad to facilitate the recognition of the filiation of the child and not to deprive him of his rights, or accepting such arrangements on their own territory.

Abortion

As we have noted, interrupting pregnancy continues to be one of the unresolved issues in bioethics and is one that generates huge confrontation between the highly polarized positions of the right to life of the unborn—'pro-life'—and the right of the pregnant woman to choose what happens to her body—'pro choice'.

The first point of disagreement concerns the issue of when we can talk of abortion. As explained, the majority view in the scientific literature and also among bioethicists is that we cannot talk of abortion if it is not intrauterine: that is, before the embryo has been implanted in the uterus. Until that point, it is not possible to speak of pregnancy or the 'individuation' of the future being with the start of organogenesis.

If we accept this view, the next question is to consider whether the value of this human life in development can be balanced with other values such as the health of the foetus, the health of the pregnant mother, which could be compromised, or the freedom of the mother to decide that she does not wish to continue with the pregnancy.

Legal regulation must address this issue of how to weigh these values. Many countries that have legislated on access to abortion do this, although they do not give the same weight to all these values. We can distinguish between the following approaches:

- Those laws that allow abortion when prenatal life enters in particular conflict with other values—health of the foetus, health of the mother, or serious damage her dignity, in the case of rape—but do not allow for a simple decision by the mother not to continue with the pregnancy, without any other factors being involved.

  Many countries permit abortion on these ethical or medical grounds but do not accept the criterion of the freedom of the pregnant woman, as there is another life at stake, one that is more vulnerable and must thus be granted greater protection.
- Those laws that establish a system of time limits, so that abortion is permitted and the role of the law is to establish these limits reflecting the stage of development of pregnancy, as the later the interruption, the more serious it is deemed to be, as the moment at which the foetus

becomes viable is approached (a foetus is considered to be viable outside the womb from week 22 to 24) (Sánchez 2014). Thus, when the reason for interrupting the pregnancy is simply the pregnant woman's decision, a shorter time limit applies, up to a maximum of 12 or 14 weeks. When the reason is medical, given the additional time needed to perform diagnostic tests, whether for foetal malformation or the impact on the health of the pregnant woman, the limits are usually longer, up to week 24 or beyond if the medical circumstances justify it.

With respect to the bioethical arguments in favor of one or other option, I refer back to the discussion above, regarding the value of incipient human life and the various positions described earlier in this chapter.

Research

Finally, with respect to the use of the embryo or of embryo-derived stem cells for research, it is important to note that such work is necessary for scientific progress. The potential of embryo-derived stem cells is not found in any other human cells, despite ongoing research to find alternatives, such as the use of IPS (induced pluripotent stem cells), resulting from cellular reprogramming. However, the results have not been very promising in the short term (Shi et al. 2017).

Progress in the understanding of diseases and possible treatments necessarily involves ongoing research using pluripotent genetic material. This can be obtained either from material specifically generated for research—the creation of embryos for research—an option that has been ruled out by most countries in order to avoid the 'commercialization' of embryos, or through use of embryo material that has been discarded for reproductive purposes or has been specifically donated for such research. In countries in which assisted reproduction is a longstanding reality, such as Spain and the United Kingdom, laws already provide for these two options and couples are invited to consider them. They are asked what use they wish to make of any embryos remaining at the end of their parental project, once they do not wish to have any more children. The options are: to donate them to research, to donate them to other couples, or for destroy these embryos. This allows for a secondary use of these embryos, beyond the couple's initial purpose.

The use of biological material to make progress in our understanding of infertility and many other issues is highly valued by the scientific community and is of great benefit to society in general and future generations, by preventing illness and disability.

Legal basis and regulatory framework

Ethical discussion of the value of the human embryo has necessarily been accompanied by analysis of the legislative frameworks that have established what is permitted and what is not at any given time or place. During the development of bioethics over the last 50 years, both the ethical debate and the legal frameworks have evolved, although this has varied from country to country.

In all of these, the starting point was the formulation of the fundamental right to life, set out in the Declaration of Human Rights of 1948, and reflected in the democratic constitutions of most states. However, the formulation 'right to life' admits of interpretations and differences of emphasis, in many cases leading to specific regulation in individual laws (e.g., assisted reproduction, abortion and in the Criminal Code), with the paradox that some countries punish abortion but permit the death penalty.

Although we do not have space here for an analysis of comparative law, we will use the example of Spain to illustrate these changes and their evolution, as this is one of the European countries with the most permissive legislation in this field.

Article 15 of the Spanish Constitution of 1978 states that: "Everyone has the right to life and to physical and moral integrity, and nobody may ever be subject to torture or inhumane or degrading punishment". Within the scope of this article, interpreted by the Constitutional Court, Spain approved the first law of assisted reproduction in 1988, and the first law decriminalizing abortion in 1985. Subsequently, updating these laws, the current Law 14/2006 on Assisted Human Reproduction Act was approved in 2006, followed by the Law 14/2007 on Biomedical Research Act of 2007, which regulates research with embryos, and the Organic Law 2/2010 on Sexual and Reproductive Health and Voluntary Interruption of Pregnancy Act of 2010.

According to the interpretation of the Spanish Constitutional Court in various rulings: "life is a reality from the start of pregnancy"... "the unborn child does not possess the fundamental right although this is a good that must be protected" ... "*in vitro* embryos cannot have a protection comparable to the intrauterine embryo" ... "the law aims to ensure that neither gametes nor embryos may be considered legally as commercially exploitable goods" (Constitutional Court Rulings 53/85 – 212/96 – 116/99) (Ollero 1995, Spanish Constitutional Court 1997, Spanish Constitutional Court 1999).

The jurisprudence of the Spanish Constitutional Court, then, makes it clear that the embryo is neither a 'person' nor a 'commercially exploitable good' but belongs, instead, to a specific category, and the law must reflect this by affording it gradual protection in light of the fact that it constitutes a form of human life that can give rise to a human being, although this requires certain conditions which do not always exist. This is the justification of the legal framework in Spain for regulating the different forms of intervention affecting the embryo, and the protection of 'dependent' human life (not yet autonomous) in contrast with the protection of the newborn child.

The key features of this regulation are:

- *Criminal Code*: regulating behaviours that are punishable with imprisonment, including injuring the foetus, genetic manipulation, the fertilization of human oocyte for non-reproductive purposes and reproductive cloning (Organic Law 10/1995 of the Penal Code).

Abortion is punished where it is practiced without the woman's consent or where consent has been given but the situations established in the law do not apply.

- Abortion on demand is permitted before week 14 of pregnancy and abortion for medical reasons (foetal malformation or risk to the life or health of the pregnant woman) up to a week 22 at the latest. Exceptionally, abortion is accepted beyond week 22 if "anomalies are detected in the fetus incompatible with life" or if "an extremely serious and incurable disease is detected in the fetus at the time of diagnosis and confirmed by a clinical committee".
- *Assisted reproduction*: permits various assisted reproduction techniques with scientific evidence, so long as the center meets quality and official authorization requirements, and with the knowledge and consent of the woman and her husband or stable partner. Also regulates access to these techniques by single women or homosexual female couples. It also covers the application of diagnostic and therapeutic techniques to the embryo, subject to certain scientific requirements, and the selection of embryos to prevent the transmission of hereditary genetic diseases or to favour the birth of a genetically compatible baby to cure a sibling. Surrogate motherhood is not permitted.
- *Biomedical research*: permits research with embryos left over from assisted reproduction, with the authorization of their progenitors, and with stem cells, and permits therapeutic cloning.

The end of human life and the decisions to be taken

Development of clinical practice at the end of life

The end of life in the context of illness and patient care, and the decision-making in such situations, have also been the focus of bioethics since the outset, with development over the years seeing the gradual consolidation of issues such as respect for the patient's wishes, preventing futile treatment, and applying palliative care to mitigate suffering, even when this includes terminal or irreversible sedation.

The first achievement was recognition of the patient as the principal decision-maker in their illness, recognizing informed consent as a mechanism to verify the patient's wish to accept treatment or other interventions affecting them. Although such recognition is not universal, it is clear that in recent decades it has become more prevalent in those countries which have developed laws in this regard. The development of bioethics in the second half of the 20th century contributed to major change in the clinical relationship, and progress in patient rights in many countries.

Some of this legislation, particularly in Europe and specifically in Spain, permits the patient to reject treatment even where this is necessary for life support. Thus, just as a patient can reject a drug or a surgical intervention, they may also reject artificial respiration or nutrition via a nasogastric tube or gastrostomy. The fact that this rejection may lead to the patient's death is not interpreted as constituting collaboration in the death by the attending health professionals but is instead interpreted as a consequence of the illness and the patient's legitimate exercise of their autonomy, which must be respected. Withdraw or non-initiation treatment merely allows death to occur by natural process, neither preventing nor intentionally causing it.

In addition to this right to reject treatment, by extension the law has allowed for the possibility of making such decisions in advance in a written document (advanced directives or "living will") providing for situations in which the individual has lost the ability to express themselves as a result of their illness. Such documents can be traced back to the 1970s in the USA, with cases such as that of Karen Quinlan establishing precedents (Fins 2010). Many European countries legislate for this eventuality, recognizing these documents as genuine expressions of the individual's wishes, which must be respected at the end of life decision making process.

At the same time, the medical community has made great progress in analyzing the futility of many actions that were assumed to be therapeutic, although in some cases these are also diagnostic, developing protocols to ensure that the therapeutic effort is appropriate and avoiding pointless actions. Technological developments in the field of medicine have made it possible to sustain biological life almost indefinitely, with the individual subject to machines that replace their vital functions, even when this often makes no sense and has the sole effect of prolonging the process of death for the patient and their family, without any possibility of recovery. There are also other areas, such as oncology, where resources are being used in a more rational and considered manner, with the emphasis on symptom control as part of a more palliative approach rather than pursuing totally ineffective treatments.

In the current European situation, with progressive ageing of the population and a high proportion of fragile patients with chronic pathologies, there is a vital need for criteria to ensure appropriate therapeutic effort and a palliative care approach.

The bioethical principle of justice is essential in such contexts, as health resources are always limited and scarce, particularly in a public health care system, and it is the duty both of professionals and of managers to make best use of the available resources, and not to waste them on actions that are of no benefit to the patient. This issue is addressed in detail another chapter of this book.

Finally, it is important to mention the use of sedation as a procedure to mitigate the ordeal of patients in critical situations where it is difficult to control symptoms that are causing great suffering. This is a common situation at the end of life, where the individual has no prospect of recovery and the only way to control their suffering is to reduce their level of consciousness with sedative drugs. This sedation sometimes must be at high intensities, and the patient's existing weakness means that death often occurs relatively quickly. This is what is called 'terminal sedation', in which the fundamental goal is not to produce the patient's death but rather to prevent suffering by controlling their symptoms.

Where sedation is properly indicated and is applied in the correct doses and perfusion rates, it is an example of good clinical practice that cannot be classified as euthanasia. For this reason, many hospitals have sedation protocols, designed to ensure correct use.

Respect for the patient's wish to reject treatment, whether at the time or through an advance directives document, the application by professionals of appropriate criteria to limit or adapt treatment, and the application of palliative or terminal sedation should not be confused with euthanasia or medically assisted suicide.

All of these actions are encompassed by good clinical practice and respect for the autonomy and dignity of the patient. Another issue that can lead to difficulties is decision-making by third parties (patient's relatives or legal representatives), when the patient does not have the capacity to accept or reject and have not advanced directives. In this scenario, the criteria for decision-making are not always in the best interest of the patient and will be the healthcare professionals who must act following good clinical practice, ensuring the principle of beneficence.

Euthanasia as the maximum exponent of the right to dispose of one's own life

As noted, patient autonomy in decision-making is legally recognized in many countries. However, this autonomy is restricted with respect to the right to dispose of one's own life in the form of a request for help to die.

Despite the examples that have been covered in the media in many countries over recent decades, the debate about euthanasia or medically assisted suicide continues to be very polarized. Very few countries have made significant moves in that direction, but their experiences are informative and show that the difficulties are primarily political and ideological rather than social.

Historical evolution

There have always been doubts about the justification of euthanasia and suicide. The different cultures and traditions that constitute the legacy of humanity have held different concepts of life, death and the manner of dying. Many of these traditions remain current, and they influence diverse sensibilities and ways of thinking, helping to understand the plurality of contemporary points of view on this question, and continuing to condition the positions of states, as reflected in legislation.

Even before the start of the Christian era (Hippocrates) (North 2012) it was argued that we were subject to the divine will, and this position was also adopted by Christianity. In this view, human beings are created by God and cannot take decisions that go against His design or against the laws of nature. Suicide is considered to constitute a form of murder and is always to be condemned, although the voluntary sacrifice of one's life is accepted as a testimony of faith (martyrs), in just wars or to save others.

With the exception of a few timid attempts by the stoics or during the Renaissance in the thought of authors such as Thomas More or Francis Bacon, who defended a different concept of the end of life in the face of suffering and irreversible illness, it was not until the contemporary period, with more pluralistic societies, that the rejection of the right to dispose of one's own life was questioned, and it was only more recently still, in the 20th century, that laws were passed establishing patient rights and the need to respect the autonomy of the individual and their wishes (Terribas Sala 2021).

Despite this, most countries continue to treat euthanasia or assisted suicide as a crime, although in many of these jurisdictions suicide itself is no longer an offense but to seek assistance is. Not even the accredited existence of a terminal illness that has reached its final stages or the fact of experiencing suffering that cannot be mitigated in other ways is admitted as an exception to the principle that any collaboration designed to hasten another person's death should be penalized. Very few countries have decriminalized euthanasia or assisted suicide.

Regulation in different countries, and the situation in Spain

There are only four European countries (Netherlands, Belgium, Luxembourg and Spain) that clearly regulate euthanasia and assisted suicide it at state level and establish it as a right of all citizens. In addition, Canada decriminalized euthanasia in 2016, following a supreme court ruling. All of these are decriminalization laws which regulate the requirements that must be satisfied for the practice of euthanasia or assisted suicide not to fall foul of the law. It is important to note that, where these requirements are not met, a crime has been committed. Therefore, it is not correct to talk about 'legalizing' euthanasia, but rather to decriminalize it under certain circumstances and subject to strict requirements. It is also worth mentioning the case of the states of Victoria and Western Victoria, in Australia, which have laws on assisted dying which regulate assisted suicide and accept euthanasia, under exceptional circumstances, when the individual is not able to self-administer drugs. In Colombia, assistance in dying is a right that has been recognized by the Constitutional Court since 1998, and this was corroborated by a ruling of the Supreme Court in 2015, although specific legislation has not been passed to regulate this. Finally, it is also worth mentioning New Zealand, which at the end of 2021 regulated euthanasia although only for cases of terminal illness with an estimated survival of no more than 6 months.

Other countries or states have developed assisted suicide laws, but not euthanasia legislation. An example of this is Switzerland, where assisted suicide is provided by private institutions that are not part of the country's health system, and one of these accepts foreign citizens. Citizens of nine states of the USA—California, Colorado, Hawaii, Jersey, Maine, Montana, Oregon, Vermont, Washington—along with the federal capital, Washington D.C., have the right to assisted suicide, although only in the event of terminal illness, with an estimated survival of no more than 6 months. Recently, another 19 states have been drafting proposals along the same lines.

In this panorama of varying legal frameworks, it is important to consider the experience of those European countries where euthanasia has been practiced for years, in particular the Netherlands, as these are the jurisdictions that have developed and applied legislation in this area, specifying the criteria to be met and applying rigorous controls. The Netherlands was the first country in the world to decriminalize euthanasia with a law that came into force in 2002, although since 1995, when the Remmelink Report verified that the practice of euthanasia existed in a clinical context and was becoming more widespread, it had been possible to report these cases as a result of a modification of the law governing death certificates. The findings of the report showed the need for a framework that provided legal security to reflect this reality. The 2002 act regulates both euthanasia and assisted suicide, for people over the age of 12 (until the age of 18, the agreement of their parents is required) who experience unbearable suffering with no prospect of improvement, and explicitly, freely and repeatedly express the wish to die and ask for help to do so. The doctor who receives this request must ensure that the patient is aware of all the treatment or palliative care options and is fully aware of what the request entails. The request for euthanasia may be contained in an advance directives document or "living will" and will be deemed valid if the euthanasia must be applied when the person has lost the capacity to express themselves. The request must be validated by another doctor, who verifies that it meets the criteria and, if the suffering experienced is of a psychological nature, a professional psychiatrist must also intervene. After performing euthanasia or assisted suicide, the doctor must inform the pathologist and submit the relevant reports. The pathologist then verifies that the criteria have been met and informs the Regional Review Commission. Where the criteria have not been met, the Public Prosecutor's Office is informed and, if applicable, the doctor may be subject to criminal prosecution. The Netherlands, then, applies a model based on trust in the doctors who attend to requests for assistance in dying, and which establishes subsequent checks by the Commission, which acts against anyone who is responsible for failing to respect all the requirements. The regional commissions (5 in total) submit an annual report to the Department of Health and the Department of Justice, and statistical data about euthanasia and assisted suicides performed in the Netherlands are published, along with other data of interest (Regional Euthanasia Review Committees 2022).

The regulatory systems in Belgium and Luxembourg are very similar to the Dutch one, with minor differences, and with 'ex post' control systems by Commissions that also report data annually. There are plenty of criticisms of these European models, although often not based on real figures, which seek to question the legitimacy of such regulation for its failure to provide enough guarantees and representing a potential open door to abuse of the most vulnerable people of society (elderly and disabled people). There is also speculation that many deaths are not reported to the commissions and of the persistence of a hidden euthanasia, which does not meet the requirements and is thus not registered. The requirement that is most frequently questioned is that of the patient's free and explicit choice, and it is argued that these people are actually forced to request euthanasia, in which case their deaths would be compassionate homicide or simply homicide. However, such assertions cannot be sustained without reliable data (Regional Euthanasia Review Committees 2018).

Spain has followed in the path of these countries, and in March 2021 approved the Organic Law 3/2021 on Euthanasia regulation. In recent decades, there have been significant changes in the social perception of euthanasia in Spain. The evolution of public opinion, with as much as 80 per cent support according to various studies, suggested that there was enough general awareness to embark upon legislative change (Ipsos 2018, Metroscopia 2019). In this context, medical professionals provide a guarantee, given their professional prestige and their social reputation, and their position is thus important when advocating such change.

The Spanish law, reflecting social acceptance and in response to demands over decades from activist groups such as the Right to Die with Dignity and others, states that "this act seeks to provide a balanced, secure, systematic, legal response to the demand from contemporary society for euthanasia" (Organic Law 3/2021 on Euthanasia regulation). It also states that, "when a person who is fully competent and free confronts a life situation which, in their view, violates their dignity and integrity, the good of life may be superseded by other goods and rights against which it must be weighed, as there is no constitutional duty to impose or protect life at all costs and against the wishes of its owner" (Organic Law 3/2021 on Euthanasia regulation). The definition of euthanasia, under which the law introduces a new individual right, is as follows: that action which produces a person's death directly and intentionally through a single and immediate cause-effect relationship, at the informed, express request, repeated over time by that person, and which occurs in the context of suffering due to incurable illness or suffering that the person experiences as unacceptable and which it has not been possible to mitigate by other means.

In general terms, the requirements established in the Spanish law are not dissimilar from those set out in the regulations we have considered in other European countries. They allow for a request to be made in the form of an advance directive document or "living will" and make explicit the right to conscientious objection of any health professionals who do not wish to be involved. Minors below the age of 18 do not currently have the right to request euthanasia.

It is worth highlighting a significant difference with respect to the control systems. In Spain, the legislation sets out a prior control model, in contrast with the 'ex post' control systems in the other European countries. This prior control requires the attending physician, after confirming compliance with all the criteria with another consulting doctor (a specialist in the principal disease affecting the patient), to submit a positive report to the Guarantees and Evaluation Commission, who will designate two members for a third verification and final approval, after which euthanasia or assisted suicide may be performed.

So, in Spain the practice of euthanasia and assisted suicide requires, before practice, three levels of control: medical doctor, medical consultant and positive report of the regional Commission.

Finally, all cases are also subject to a last review, after to be applied, by the Guarantees and Evaluation Commission, based on the reported documentation from the medical doctors.

This Commission is regional for each of the autonomous regions in Spain and must compile an annual report of its activity.

Previous legislation in other European countries (the Netherlands, Belgium, Luxembourg) has established an 'ex post' control system, delegating responsibility for practicing euthanasia and compliance with all the criteria to the attending physician and another doctor who provides a second opinion, in an exercise of trust in professionals that avoids the administrative bureaucratization that can prolong the patient's suffering. In contrast, the Spanish legislators deemed it necessarily to establish a 'prior control' system in order to provide both health professionals and society with greater legal security that abuses would not be committed, and euthanasia would not be practiced incorrectly. Although this model has the benefit of providing stronger guarantees for the citizens and healthcare professionals, in practice it makes the procedure slower and administratively more complex for the patient (Terribas Sala 2022).

We will need to wait before we can analyze how the law is being applied in Spain, and we will also see whether other countries take a similar approach, such as Portugal, which already has draft legislation.

References cited

Abel, F. 2012. Francesc Abel y la bioética, un legado para la vida. Proteus, Barcelona.

Albert, M. 2017. La explotación reproductiva de mujeres y el mito de la reproducción altruista: una mirada global al fenómeno de la gestación por sustitución. Cuadernos de Bioética. 28: 177–197.

Casado, M., I. Lecuona, F. Estévez Abad, F. García López, J. Martínez Montauti and M.J. López Baroni. 2021. Manual de bioética laica (II). Cuestiones de salud y biotecnología. Aranzadi, Navarra.

Congregation for the Doctrine of the Faith. 1987. Donum Vitae Instruction. https://www.vatican.va/roman_curia/congregations/cfaith/documents/rc_con_cfaith_doc_19870222_respect-for-human-life_sp.html.

Congregation for the Doctrine of the Faith. 2008. Dignitas Personae Instruction. https://www.vatican.va/roman_curia/congregations/cfaith/documents/rc_con_cfaith_doc_20081208_dignitas-personae_sp.html.

Council of Europe. 1997. European Convention on Biomedicine and Human Rights. Council of Europe, Strasbourg.

Fins, J.J. 2010. Minds apart: Severe brain injury, citizenship, and civil rights. Law and Neuroscience. Current Legal Issues 13: 367–384.

Institute Borja of Bioethics. 2009. Consideraciones sobre el embrión humano. Bioètica & Debat. 15: 2–11.
IPSOS. 2018. El 85% de los españoles a favor de regularizar la eutanasia. https://www.ipsos.com/es-es/el-85-de-los-espanoles-favor-de-regularizar-la-eutanasia.
Junta de Andalucía. 2008. Ethics and Death with Dignity. Junta de Andalucía. Consejería de Salud, Granada.
Metroscopia. 2019. Muerte digna. https://metroscopia.org/muerte-digna/.
Mujer, Madre y Profesional de Profesionales por la Ética. 2015. Vientres de alquiler. Maternidad subrogada. Profesionales por la Ética, Madrid.
North, M. 2012. The Hippocratic Oath. History of Medicine Division. National Library of Medicine. National Institutes of Health, Betsheda.
Ollero Tassara, A. 1995. Todos tienen derecho a la vida. ¿Hacia un concepto constitucional de persona? pp. 341–364. *In*: Ballesteros Llompart, J., E. Fernández Ruiz-Gálvez and A.L. Martínez-Pujalte (eds.). Justicia, solidaridad, paz: estudios en Homenaje al Profesor José María Rojo. Sanz.Universitat de Valencia, Valencia, Spain.
Organic Law 10/1995 of the Penal Code, November 23rd 1995. Boletín Oficial del Estado. 281: 33987–34058.
Organic Law 14/2007 on Biomedical Research, July 3rd 2007. Boletín Oficial del Estado. 159: 28826–28848.
Organic Law 2/2010 on Sexual and Reproductive Health and Voluntary Interruption of Pregnancy, March 3rd 2010. Boletín Oficial del Estado. 55: 21001–21014.
Organic Law 14/2016 on Assisted Human Reproduction Techniques, May 27th 2006. Boletín Oficial del Estado. 126: 19947–19956.
Organic Law 3/2021 of Euthanasia Regulation, March 24th 2021. Boletín Oficial del Estado. 72: 34037–34049.
Regional Euthanasia Review Committees. 2018. Annual Report. Regional Euthanasia Review Committees, Rotterdam.
Sánchez Luna, M. 2014. Límite de la viabilidad en la actualidad. Anales de Pediatría. 80: 346–347.
Shi, Y., H. Inoue, J. Wu and S. Yamanaka. 2017. Induced pluripotent stem cell technology: a decade of progress. Nature Reviews Drug Discovery 16: 115–130.
Terribas Sala, N. 2014. Debate sobre el aborto. Desclée de Brower, Bilbao.
Terribas Sala, N. 2021. Eutanasia como alternativa. pp. 273–285. *In*: Beca, J.P. and R. Armas (eds.). El final de la vida. Mediterráneo, Santiago de Chile, Chile.
Terribas Sala, N. 2022. Ley Orgánica de regulación de la eutanasia en España: cuestiones polémicas sobre su aplicación. Folia Humanística. 7: 1–25.
The Holy See. 1992. Catechism of the Catholic Church. https://www.vatican.va/archive/catechism_sp/index_sp.html.
Torres Quiroga, M.A. 2019. Maternidad y gestación en venta. Fabricar bebés en la era neoliberal. Universitat Barcelona, Barcelona.
Spanish Constitution, 29th December 1978. 1978. Boletín Oficial del Estado. 311: 29313–29424.
Spanish Constitutional Court. 1997. Sentencia TC 212/1996 de 29 de diciembre. Boletín Oficial del Estado. 19: 1–26.
Spanish Constitutional Court. 1999. Sentencia TC 116/1999 de 17 de junio de 1999. Boletín Oficial del Estado. 162: 1–29.
UNESCO. 2005. Universal Declaration on Bioethics and Human Rights. UNESCO, Paris.

Chapter 6
Ethics of Health Care Allocation of Resources. The Case of Organ Transplantation

*Marius Morlans Molina** and *Marc Antoni Broggi Trias*

The ethics of the allocation, prioritization and rationing of health care resources

Introduction

In moral philosophy, distributive justice is concerned with the reasoning and criteria that govern the allocation and prioritization of basic resources and goods. Criteria such as need, merit, purchasing power and the optimization of outcomes are grounded in different understandings of distributive justice. This section reviews the foundations and principles of the different schools of thought—utilitarianism, egalitarianism, libertarianism and communitarianism—with reference to the key figures. It then provides a series of examples to illustrate their practical implications.

Theories of distributive justice

Distributive justice deals with the distribution of basic goods and resources in a society in line with certain ethical principles. The concept

Comitè de Bioètica de Catalunya, Spain.
* Corresponding author: 7234mmm@comb.cat

dates back to Aristotle (1985), whose formulation, "if the persons are not equal, they will not have equal shares" defines justice as the problem of treating equals as equals and unequals as unequals. However, this raises the question of just what makes people equal or unequal: is it need, effort, personal contributions to society or individual merit? The different theories of distributive justice are defined by the way they answer this question. This section reviews the main theories, examining the principles in which they are grounded and their criteria for allocation, with reference to the key figures in each school of thought.

Utilitarianism

Utilitarianism is an ethics of consequences that judges the goodness of decisions and actions by their outcomes in abstraction from individual motives and intentions. It originated in the United Kingdom in the 19th century and its founding figures included Jeremy Bentham. Utilitarianism is based on the premise that well-being—understood as happiness—is the ultimate goal of human beings. Consequently, the primary motivations for utilitarianism are the pursuit of pleasure and the alleviation of suffering. The ethical principle of utility consists in maximizing the happiness of the greatest number of people, treating individual interests as equal: "everybody to count for one, nobody for more than one" (Bentham 1987).

For John Stuart Mill, good actions and thus good policies are those that provide the greatest well-being to the most people. In contrast to his mentor, Bentham, Mill (1987) prioritized the cultural and spiritual components of happiness above more worldly pleasures. He distinguishes between what is "useful" and what is "expedient": the former contributes to general happiness, while the latter relates to the pursuit of a specific or even personal goal. Mill's approach thus seeks to balance utilitarianism in public life with the moral traditions of the private realm.

In *On Liberty*, published in 1859 (Mill 1982), he expresses his concern for the "tyranny of the majority", which is exercised through the mechanics of voting and fails to account for the rights of minorities and individuals. Mill is a staunch defendant of individual liberty, understood as self-determination: "over himself, over his own body and mind, the individual is sovereign" (Mill 1982). Any intervention that impinges on individual liberty, he argues, can only be legitimate when it seeks to prevent the individual from causing harm to others through exercising this liberty. His utilitarianism is thus tempered by his liberal principles.

Utilitarianism was the defining feature of an era of public policy on resource distribution. For health care, it is the cornerstone of the concept of the quality-adjusted life year (QALY), a major contribution to measuring the outcomes of health care procedures. Originally proposed by Williams

(1985), the QALY approach seeks to measure the outcome of procedures by combining years of survival and quality of life, using appropriate scales or indexes. It allows the objective prioritization of medical procedures based on an analysis of their cost-effectiveness. Efficiency is the principal criterion for the allocation and prioritization of resources, reflecting the utilitarian concept of distributive justice.

The main strength of this approach is that it provides the information needed to prioritize and allocate resources based on objective criteria, which requires rigorous and accurate measurement methods. However, it reduces people's value to a quantifiable estimate of their health or the expectations of the success of the procedure in question (Puyol 1999). This leaves it open to the criticism that impartial does not mean impersonal, since different people have different needs. Its use for prioritizing individuals raises a number of questions, since the benefit for the majority may come at the cost of excluding a marginalized minority. The limit of efficiency is the respect for human rights.

Egalitarianism

Egalitarianism is inspired by the ethical principle of equality. Everyone has the right to be treated as equal because we all have an intrinsic value that makes us unique. This value—our dignity—differentiates us from things, which can be exchanged and thus have a price (Kant 2005). The principle of equality states that people are entitled to have their basic needs met, including health. Health care is essential to restore the normal functioning of humans when we fall ill. Without this functioning, we cannot aspire to enjoy freedoms, basic goods or social rights (Puyol 2001). The right to health care is one of the welfare state's greatest achievements.

John Rawls is one of the founding figures of the egalitarian concept of justice and is among the most influential moral philosophers of the 20th century. Rawls (1971) draws on ideas from Kant and contractualism as the basis of his critique of utilitarian justice. He defines justice in terms of equity, which takes precedence over the good. His distinction between the rational and the reasonable echoes the Kantian distinction between the hypothetical imperative and the categorical imperative. Rational approaches to the principles of justice must be reasonable to the parties involved in an "original situation", who, understandably, defend the interests of those they represent.

Rawls' work leads us back to the idea of the social contract and to enlightenment figures like Hobbes, Locke and Rousseau. His point of departure is a hypothetical "original position", where individuals act under the "veil of ignorance", without knowing what will happen to them in the future or the situation they will end up in. Under these conditions, individuals adopt the "maximin" strategy, which has its roots in game theory: not knowing if they will win, each player seeks to ensure their worst outcome will be as good as possible, minimizing their maximum losses and maximizing their minimum gains.

The individuals of Rawls' original position thus agree two principles of justice. First, everyone is equal when it comes to demanding an adequate social model of basic rights and freedoms for all, which guarantees the equitable value of equal political freedoms. Second, social and economic inequalities must satisfy two conditions: they must first be linked to roles and positions to which everyone can aspire under conditions of equal opportunity; they must also promote the greatest benefit for the least-advantaged members of society. This last condition is called the "difference principle" and is fundamental for correcting the impact of social differences on well-being (Rawls 1971).

Rawls has attracted criticism from some egalitarians. Dworkin (1981), whose ethical perfectionism brings his position close to communitarianism, criticizes the absence of values and principles other than equality and freedom among the individuals of Rawls' original position. He proposes providing everyone with the same basic resources or goods to guarantee their personal development and well-being. In defence of Rawls, individual equality and freedom can be conceived as basic requirements for personal development, respecting free determination and unhindered by a universal conception of the goals of individuals.

Amartya Sen (2009), on the other hand, distinguishes between equality of opportunities and equality of capabilities, understood as the real possibility to exercise basic functions by converting resources into utilities or well-being. Sen draws inspiration from the work of Marta Nussbaum (1992), who proposed 10 central capabilities. Capability theory allows equity to be extended to the sociocultural factors that condition effective access to the health system.

However, Sen's main contribution is his critique of the transcendental institutionalism represented by Rawls. Philosophical conceptions of justice centred on institutions and procedures, he argues, are ideal constructions with serious limitations when it comes to resolving major injustices in the real world. Instead, Sen proposes analysing and comparing the most just alternative for a given situation. His work is influenced by Kenneth Arrow's social choice theory and the Indian tradition of *niti*, which emphasizes social interventions that seek to denounce, avoid or correct the most visible injustices (Sen 2009). This approach allows injustices to be addressed in countries that do not share the political tradition of Western democracies and, in this sense, Sen's ideas can be seen as complementing those of Rawls.

Libertarianism

The term libertarianism was coined to distinguish this extreme school of thought from traditional liberalism, regarded as more progressive. The cornerstone of libertarianism is individual liberty and the obligation not to interfere in other people's lives. Its main figures include Nozick (1974), who argues that there are only people with their individual lives and there is no such thing as the common good. For Nozick, justice consists of fiercely defending the right to private property and the State's functions should thus be limited to this defence. Access to health care depends on the purchasing power of individuals and their ability to pay for health insurance, whose prices are set by the market. Nozick is opposed to the right to publicly funded health care, since this means forcing citizens to pay taxes. Inequalities in health care and access to it are regarded as unfortunate but not unjust. Charity may be exercised towards others, provided it is a free and individual choice. However, there is no obligation to provide people with care (Nozick 1974).

The main criticism of this theory is that it fails to recognize the common good and thus individual health as a public good to be protected. It undermines public health policies that are effective for maintaining and promoting community health (Nozick 1974). Leaving regulation of access to health care to the market has been shown to be inefficient. Despite some countries spending a high percentage of GDP on medical care, a large proportion of their population lack coverage, since they cannot afford health insurance.

Buchanan, another key figure of libertarianism and an economist-cum-philosopher, attacks the welfare state as an economic and moral aberration. Inequality is a fact of life and the State must not try to compensate for this with social policies. Buchanan (1986) argues that redistributive social policies are absolutely inefficient from an economic point of view. From a moral standpoint, State intervention represents intolerable interference on individual freedoms. He thus proposes a new social contract that sets ethical limits on taxation, such that the level of taxes imposed on the richest is no more than they would be willing to pay in a society without poor people. Morality is grounded in the predisposition of individuals to make sacrifices for the common good or in the general interest. However, this must not be imposed by force (Buchanan 1986).

In the field of bioethics, Engelhardt is representative of the libertarian school of thought. His work explores the difficulty of reaching agreements on ethical issues among what he calls "moral strangers", people who do not share the same moral system. The plurality of worldviews and beliefs that characterize postmodernity leads him to propose a secular morality based on certain minimums that allow consensus to be reached among people with different ethical values. For Engelhardt (1986), the authority of measures that involve others in a secular and plural society is grounded in the principle of permission. In contrast to the principle of autonomy, this principle does not emphasize autonomy and personal freedom. It allows for the fact that, in certain cultures, patients and their relatives agree that professionals, exercising a trust-based paternalism, act for their benefit, without requiring the procedure of informed consent, whose complexity is beyond their understanding.

According to this principle, the right to health care is contingent on the permission of others, just as people cannot be made to work and their property cannot be confiscated. The right to care is dependent on having an insurance policy, which defines the scope of the services covered. Engelhardt does not believe it is possible to reach a consensus on a universal theory of justice and instead argues that solidarity and charity should be used to address social inequalities (Engelhardt 1986).

Communitarianism

Alasdair MacIntyre is one of the leading figures of communitarianism, which is based on a critique of the current language of morality. Debates on morality, he argues, draw on fragments from discourses whose contexts have moved on. As a result, both the theoretical and practical understanding of morality has been partially lost. By default, everyday life has become characterized by a culture that favours emotional responses to moral problems over rational ones, albeit without being fully conscious of this situation.

Another contribution from MacIntyre's work is the idea that choosing the narrative is itself a procedure. Narrating a philosophical history is different from the history of philosophy: the narrative interpretation obviates the distinction between matters of fact and matters of value, which is fundamental to the academic world. MacIntyre's work is a critique of the discourses of academic disciplines that use the empirical base on which they were founded to stake the claim to an axiological neutrality—dubious at best—while overlooking the fact that the transmission of the body of the discipline's doctrine is a narrative act in its own right and one that is not exempt from valuations.

He proposes an ethics based on the virtues that guide our pursuit of the goods that characterize a full life. Yet goods and virtues can only be discovered and cultivated in personal relations established in the bosom of communities, bound together by the shared vision and understanding of these goods. To isolate oneself from the community is to condemn oneself to moral solipsism, he argues, citing Nietzsche—the counterpart to Aristotle as the paradigm of an ethics of virtues—as a case in point (MacIntyre 1981).

MacIntyre (1981) analyses the dispute on justice between Rawls and Nozick to illustrate his argument. Justice, he claims, is the social virtue that maintains the cohesion of communities and he frames the disagreement between the two thinkers, both members of the same community, as proof that lack of context has made it impossible to debate morality. MacIntyre (1981) criticizes the concepts of justice of both Rawls, based on the criterion of equality, and Nozick, based on the right to private property, arguing that neither correctly contextualizes the problem and their arguments fail to account for the past and the tradition of the community. Moreover, their considerations of what is just or unjust ignore the concept of merit, a notion that only makes sense in the context of communities, which are essentially bound by consensus on what is best for both the individual and the community. Full civilized, moral and intellectual life, in which the allocation of resources is based on merit, understood as the individual's contribution to society, is only possible in such communities.

However, while MacIntyre's diagnosis of the problem may be accurate, his alternative is questionable. Far from providing a solution, his idea of communitarianism is part of the problem. One of the biggest challenges to ethical behaviour today is the possibility of moral dialogue between people from different cultures. While a return to local communities grounded in their own traditions may make their members happy, this does not foster understanding among members of communities with different moral traditions.

Michael Sandel, another communitarian thinker, offers a forceful critique of meritocracy, arguing that it fails to provide the social mobility needed to increase equality. He argues that biases inherent in the selection process for university education based on merit and equal opportunities tend to reproduce existing social differences, largely favouring access among children from higher-income families. This situation has widened the gap between graduates and people employed in working-class jobs, reinforcing the self-esteem of the "winners", who attribute their merit to personal sacrifice without taking into account the advantageous social position from which they started out. The arrogance of the winners and a tendency to look down on manual labour is one of the factors behind the rise of populism, a phenomenon that is more pronounced in the United States than in Europe (Sandel 2020).

From theory to practice: a few examples

This section looks at four examples of the theories covered in the first part of this chapter: the work of Diego Gracia; the state of Oregon's pioneering health care prioritization initiative; public involvement in selecting the criteria to prioritize waiting lists; and the handling of medical emergencies. All these examples show the challenges of applying different theories of justice in the real world. While theories do not solve problems, they can help us to understand the concepts involved and to identify the underlying principles and values of the criteria used to determine priorities (Churchill 1992). It is people, with their prudence and experience, who provide solutions, and choice of the corresponding criteria is guided by their ideas about good moral behaviour.

Diego Gracia's priorities in Health Care

In an influential text, Diego Gracia (1989) identified four concepts of justice: as virtue or personal justification; as a value that sets a moral example; as a rational blueprint of rights and responsibilities; and as real experience. In 1995, at a meeting of the European Association of Centres of Medical Ethics, organized by the Borja Institute of Bioethics (*Institut Borja de Bioètica*) and chaired by Francesc Abel, Gracia complemented his idea of distributive justice with three criteria for the allocation of health care resources (Gracia 1996).

The first is that health is a personal and private matter. Each individual has an idea regarding their own health, depending on their lifestyle and values. Health is about how our bodies enable us to pursue our personal life projects. The distribution of health care resources cannot be addressed without taking into account that this is largely done by individuals, not the State. To attribute all care to the latter is to expropriate the body and life and, paradoxically, to promote illness.

The second criterion is that public responsibility lies in ensuring everyone has access to quality basic health care. Under the principle of equity, which guarantees equal opportunities, the State must cover treatment of health problems, but only those that prevent the normal functioning of the human body. Restoring or improving health is necessary to ensure equality in the exercising of the rights and responsibilities inherent to being a citizen.

The third criterion proposed by Gracia (1996) is that the scarcity of resources means it is necessary to maximize good and minimize harm. When a lack of resources constrains the duty to help everyone equally, an exception must be made to the principle of equity and the criteria for prioritization must be debated. This means weighing up consequences and choosing the alternative that delivers the greatest health benefit and does the least harm.

Gracia's (1989, 1996) approach is based on a pragmatic mixture of different theories of justice. His first criterion is libertarian, highlighting the value of personal autonomy in looking after our health. The inspiration behind the second is clearly egalitarian, since it recognizes the universal right to publicly funded care. Lastly, in contrast to the deontological nature of the first two criteria, the third—with its emphasis on maximizing good and minimizing harm—is utilitarian, based on an ethics of consequences.

The Oregon state health plan

The Oregon state health plan was a pioneering initiative to prioritize access to health care services by analysing the cost-effectiveness of treatment (Benach and Alonso 1995). In 1989, Oregon undertook a project to extend its health care coverage to the state's least-favoured residents through a series of policy objectives that sought to guarantee a basic package of services for everyone. The initiative began by selecting 1,600 health diagnoses and conditions, together with their respective treatments or services.

The expected benefit of treatment was calculated for each condition and treatment, alongside the corresponding duration and costs. The long-term cost–benefit ratio was used to rank procedures by efficiency. The expected benefits were calculated using the Quality of Well-Being scale, which measures 43 states of functional capability, ranging from full health (1) to death (0). The cost of services and treatments was calculated based on the records of insurance companies and Medicaid, the subsidized federal health care scheme in the United States.

The procedure underwent numerous modifications in response to criticism and opposition from the federal government. Seventeen categories of treatment benefits were identified and each diagnosis—treatment pair was assigned to a category. The Delphi method was then used to score them from 0 to 10, with input from the public. This resulted in the inclusion of prevention programmes, such as family planning and maternal and child health. After a number of final amendments, a list of 696 services was approved, 565 of which were included in the basic package.

The initiative was widely debated and criticized, both academically and on political and ideological grounds. Its method has many shortcomings, not least that the efficacy of most treatments was unproven, since this would have required long-term research using clinical trials (Kaplan 1993). Moreover, quality of life varies over time and the value depends on life expectancy. This intersubjective variability makes it hard to reflect in a mathematical formula (Calsamiglia 1996).

From an ideological perspective, the initiative was also criticized as an exercise in rationing that only affected the poor. This criticism would be valid if all the health care needs of the population had been met prior to the change. However, this was not the case. The plan made clear the limitation of resources from the outset and it was in this context that it sought to guarantee the basic minimum of quality care proposed by Gracia (1996). Analysis of cost-effectiveness generate data on the efficiency of health care procedures and support health policy decisions that could limit access to specific treatments for at-risk groups due to the lack of expected benefit and high costs. However, applying decisions of this type to the clinical relationship could undermine trust in health care workers, forcing them to choose between defending the interests of patients or prioritizing the efficiency of the system (Morlans 1996).

Managing waiting lists for elective surgery

In Spain, the Spanish Network of Agencies for the Evaluation of Technology and Services of the National Health System (*Red Española de Agencias de Evaluación de Tecnologías y Prestaciones del Sistema Nacional de Salud,* RedETS) is responsible for evaluating new health care technology and procedures. The network has created an online application (web.pritectools.es) that prioritizes technology using 17 weighted criteria grouped under five categories: illness, outcomes, economic impact, implementation and availability (Espallargues 2018).

One of the objectives of public health systems is to guarantee equitable access to waiting lists (Edwards 1999, Hadorn 2000, MacCormick et al. 2003). The Catalan Health Information, Evaluation and Quality Agency (*Agència de Qualitat i Avaluació Sanitàries de Catalunya,* AQuAS), which is part of RedETS, has developed a system for prioritizing cataract operations and hip and knee replacements based on preferences chosen by the public (Sampietro and Espallargues 2004, Tebé et al. 2015). The method has two steps. First, a group of people directly linked to the process select the criteria they believe to be most important. Second, a wider group of randomly selected individuals are interviewed and asked to rank them. Multivariate regression is then used to weight the criteria.

Pain is the top criterion for replacement operations, followed by inability to function and severity, then, with a lower weighting, inability to work, not having caregiver, being a caregiver, and capacity for recovery. For cataracts, sharpness of vision replaces pain and severity. A score from 0 to 100 is calculated for each patient using the weighting. They are then listed in descending order, without taking waiting times into account. The public prioritized clinical criteria, such as pain, severity or sharpness of vision, followed by criteria related to need, such being or needing a caregiver. This reflects an egalitarian approach to justice. Criteria related to efficiency, such as prioritizing people who are off work and those with greater potential for recovery, were lower down the ranking.

These pioneering experiences were used to develop a procedure for prioritizing waiting lists for elective surgery in the Spanish health system (Allepuz et al. 2009, Solans-Domenech et al. 2013). Not only does determining and communicating criteria draw attention to limitations on resources, it also addresses them transparently, avoiding arbitrary decisions that could exclude vulnerable groups like the elderly. Public participation helps to build trust in health services and to ensure their rational use.

Managing medical emergencies

When dealing with medical emergencies, it is important to distinguish between three concepts related to clinical decisions. These concepts, which are often confused in practice, are prioritization, adequacy and limitation of treatment, and rationing (Martín-Fumado et al. 2020).

Prioritization means establishing an order in which patients receive care based on predetermined criteria. Everyone receives adequate treatment but not at the same time. Triage in medical emergencies is an example of a prioritization procedure that involves making quick decisions and accurately classifying illness and trauma in order to deal with the most serious cases first (Bazyard et al. 2020).

Ensuring the adequacy of treatment means avoiding futile intervention in cases of advanced and irreversible illness. It aims to avoid harm caused by therapeutic obstinacy, promoting the adequate use of resources in a context without explicit limits. The decision to withdraw life support treatments in intensive care is based on the chances of survival. This is evaluated using scales or indicators that quantify the functional state of organ systems, such as the Sequential Organ Failure Assessment (SOFA) (García Lizana et al. 2010) and the Acute Physiology and Chronic Health Evaluation II (APACHE II) (Gonzalez-Castro et al. 2017).

Lastly, rationing occurs when there is a scarcity of essential resources. It involves subjecting them to a distribution established by the authorities, since resources can only be allocated to one person. The COVID-19 pandemic highlighted the urgent need for criteria for rationing treatments. One example of such criteria is the bioethics approach developed by Emanuel et al. (2020) based on four fundamental values: (1) maximizing benefit, understood as preserving life expectancy as much as possible; (2) treating everyone equally: avoiding, for example, prioritization by order of arrival; (3) promoting instrumental values, such as by prioritizing the protection of essential workers; (4) prioritizing the most serious cases or the people at highest risk.

These four criteria, which seek to maximize the benefit for the majority (measured in years of life gained in the short and long term based on life expectancy prior to illness) are based on previous outcomes for specific groups. The criteria aim to provide objective data to support decision-making. Yet while they bound the level of uncertainty, they cannot predict individual outcomes. Since exceptional rationing calls into question basic features of public health systems like the principles of universal access and equal treatment, they must be publicly agreed and approved. Prioritization processes that involve public participation, such as those discussed above (Sampietro and Espallargues 2004), show that people also value non-clinical criteria. Moreover, they reflect the prevailing values in society and give legitimacy to the criteria and decisions of the health authorities.

Ethical problems of organ transplantation

Introduction

The scientific approach of academic education and its focus on effectiveness leave little room for moral doubt. Nevertheless, organ transplantation shows how the difficulty of making decisions is not scientific or technical in nature (due to lack of knowledge or skills) but is moral (either in terms of moral concern or, more broadly, as a conflict of values). In the case of brain death—a basic concept in organ procurement—the question arises as to whether explicit consent from the donor is required. The allocation of organs to patients on a waiting list is an example of prioritizing scarce resources. In addition, transplants from living donors, a less common source of organs, call into question the very foundations of the clinical relationship.

The controversies surrounding brain death

In an article describing a group of patients in deep non-reactive coma, kept alive using mechanical ventilation, the authors end with a question: "where to place that fraction of a second that separates life and death?" (Mollaret and Goulon 1959:15). An answer was provided 10 years later by a committee of experts at Harvard University, which defined the concept of brain death and identified the criteria to diagnose (Ad Hoc Committee 1968). The concept has been controversial from the outset. Hans Jonas (1984) argued it is not possible to narrow down the process of dying to a specific moment. The philosopher rebuked scientists for daring to enter into a realm that had hitherto been reserved for philosophical speculation, especially for a matter as pragmatic as the procurement of organs for transplantation. The controversial question of whether brain death is the same as biological death has recently been revisited (Miller et al. 2021).

Death is a process of biological decomposition that continues after cardiopulmonary arrest, as shown by the subsequent growth of nails and hair. Total injury of the brain stem suppresses the blood circulation automatisms and respiratory movements, eventually causing the heart and lungs to stop, regardless of life support treatments. As a result, diagnosis of the destruction of the brain stem confirms irreversibility of the process of death. This is in contrast to patients in a persistent vegetative state, where the cerebral cortex is destroyed but the brain stem remains intact.

Under these circumstances, patients can continue living, provided they receive hygiene care and artificial nutrition, but they cannot relate to the world around them because they lack cognitive functions (Multi-Society Task Force on PVS 1994).

The concept of brain stem destruction syndrome is a clearer alternative to brain death, since diagnosis allows certification of death by confirming the irreversibility of the process. This means life support treatment must be withdrawn from all patients diagnosed with this syndrome, regardless of whether they are organ donors, in order to ensure the efficient use of intensive care beds. It also helps us to understand the clinical relevance of the syndrome, in isolation from the issue of organ procurement.

Organ donation: opt in or opt out?

Legislation governing consent for organ donation varies from country to country. Most countries require the donor's explicit consent, although a small number assume donation as the default position unless the individual has opted out. Spain is among these countries and has the highest donation rate in the world. This raises the question of to what extent this is a product of its policy of implied consent (Willis and Quigley 2014).

In contrast, in the United Kingdom, where donations are far below the required level, there has been talk of changing the legislation governing donation, despite evidence that the decision by Wales to adopt an opt-out system has not increased donation levels (Rudge 2016). Indeed, the Spanish health authorities have themselves challenged whether implied consent is behind the country's high donation rate, citing, among other factors, respect for the entitlement of relatives to express the will of the deceased when it is not in writing (Fabre et al. 2010).

From an ethical perspective, both options highlight the tension between two values: individual freedom and the common good. In clinical practice, bioethics has enshrined consent as the means to ensure individuals are respected. This is based on the premise of people's autonomy and ability to freely decide on health matters, given sufficient information. This is one of the main legacies of the Belmont Report (National Commission 1978).

However, justice also means promoting the most efficient procedures that deliver the greatest health at the lowest cost. If the outcomes of a kidney transplant are measured in terms of life expectancy and quality of life (QALY) (Williams 1985), the annual cost of a transplant from a deceased donor is a quarter of the cost of dialysis, the most widespread and costly treatment for advanced kidney disease (Arredondo et al. 1998). These savings can be used to meet other care needs that benefit more people, with a clear collective benefit.

All this raises the question of whether limitations can be applied to individual freedom in the name of the common good. There are many examples of such legislation, including road safety, the prohibition and limitation of tobacco and alcohol, and the requirement to declare certain infectious diseases. Different arguments apply in each of these cases. However, when it comes to health care, the consent procedure has become firmly embedded as the means to ensure people suffering from illnesses and who are already vulnerable and lacking autonomy are able to decide for themselves. This is why the procedure should be respected for organ or tissue donation and should be taken as the last will of an altruistic and charitable individual.

The allocation of organs from deceased donors

The procurement of organs for transplantation from deceased donors is highly complex. Maximizing the social benefit from an altruistic individual decision raises the issue not just of the ownership of organs but—perhaps most importantly—their allocation. As a scarce good whose demand outstrips supply, the ethical, legal and social issues for the parties involved have provoked extensive debate, giving rise to a number of recommendations (Strathern et al. 2011).

Specific state agencies are responsible for the promotion, procurement, transport and allocation of organs. Spain's National Transplant Organization (*Organización Nacional de Trasplantes de España*) is among the most successful examples in the world (Morlans 2019b). The organization grew out of the transplant coordination unit established at Barcelona's Vall d'Hebron hospital in 1984 (Valls 2009).

Organs are allocated based on the criterion of urgency, with priority given to paediatric cases (if the size of the organ permits) and taking into account ABO blood group compatibility. The cold ischaemia time of the extracted organ places limits on the distance and the facility where the transplant can take place. Patients on waiting lists are classified by severity, except for kidney transplants. Severity is estimated based on life expectancy, evaluated in terms of the resources required to keep the patient alive (including extracorporeal membrane oxygenation, for cases of advanced heart disease, or mechanical ventilation, for patients with serious respiratory problems). For liver transplants, the Model for End-Stage Liver Disease (MELD) score is used for patients over 12 years of age, with the Paediatric End-Stage Liver Disease (PELD) model used for younger patients. These scores are calculated based on biochemical data that reflect the functioning of the liver and other biological variables. The criteria and the procedures for their application are the product of scientific consensus. They are also public and can be reviewed, allowing results to be audited (OPTN 2022).

The scoring procedure for kidney transplants aims to guarantee: (1) the transparent allocation of the organ; (2) equity to ensure patients with similar characteristics have the same opportunity, regardless of how long they have been on the waiting list; and (3) the benefit to the patient and not the transplantation centre. The calculation takes into account time spent on dialysis, donor risk factors, HLA compatibility, antibody levels, the age difference between the donor and the recipient, and the distance between the facility of origin and the site where the transplant will take place (NHSBT 2019).

Waiting lists for kidney transplants could be prioritized using the same method for elective surgery, based on public participation. This would make it possible to complement clinical criteria by identifying and attaching a weighting to socially valued criteria, such as functional capacity, the need for a caregiver, occupational status and family responsibility, which complement clinical criteria (Espallargues 2018). In addition to identifying and prioritizing the most vulnerable people and those in greatest need, the criteria build trust in the health system and can encourage organ donation (Morlans 2019c).

Living donors: a clinical and moral challenge

While there have always been transplants from living donors, longer waiting lists have seen an increase in the practice. Extracting an organ from a living donor violates the ethical principles of beneficence and non-maleficence, the foundation of any clinical relationship. We are no longer just dealing with restoring the health of an individual: the whole or partial extraction of an organ means that donors, who subject themselves to a form of mutilation in undergoing the procedure, must be in good health (Ingelfinger 2005).

The ethical justifications of the atypical clinical relationship established in cases involving living donors include respect for the donor's decision and the surgeon's commitment to limit the harm and risks inherent to extracting the organ. By consenting, donors assume the lack of a benefit and accept the loss and risks of the procedure, on the condition that these are minimized. The principle of autonomy prevails over beneficence and non-maleficence (Truog 2005).

There are two main threats for living donation: coercion and payment. Donation by conviction, that is, the reasonable and free decision of an autonomous individual, stands in sharp contrast to donation due to coercion, when external action constrains or overrides the donor's will. Examples of extreme coercion include people who are deprived of liberty, either because they have been imprisoned by the State or because they have been kidnapped. However, more subtle forms also exist, such as the coercion of relatives in positions of economic or emotional dependence.

Any coercion that goes beyond reasonably convincing an individual means the decision is not free and thus invalidates consent. Consent can only be valid if sufficient information is available and individuals are free and fully able to decide (Simón 2000). The surgeon is responsible for ensuring consent is correctly obtained. Trust in the surgeon and the confidentiality of the clinical relationship can help bring to light cases of coercion. It must be clearly explained to the potential donors that they can be rejected as unsuitable without any need to state a reason (Morlans 2011). Under Spanish legislation, the validity of consent is certified by ethics committees (Royal Decree 2070/1999).

A second threat to transplants from living donors is material remuneration: that is, the sale of organs. In contrast to coercion, while the material interest or economic benefit behind the donation conditions the will of the donor, it does not constrain it. People who donate for an economic benefit or interest are as free as those who do so for more altruistic reasons. Arguments about the integrity of the body and the indignity of the sale come up against the donor's clear desire to use their body as they see fit (Richards 1992).

While donations for economic benefit do not violate the principles of autonomy or consent, respect for the principle of justice is less clear. Justice is the principle that regulates the exercise of freedom, respecting equality. It seeks to prevent harm to third parties in the name of individual freedom and is fundamental to the collective right to decide on individual behaviour when it goes beyond the private sphere (Morlans 2019a).

Libertarian societies in which access to medical care depends on individual purchasing power are more likely to accept donors' decisions, regardless of their motives. However, in egalitarian societies with publicly funded health systems, while donation for material gain may be a free decision, it is not a just one. The fact that the only participants in this practice are the least-favoured and most vulnerable members of society mean it is inherently inequitable (Morlans 2011).

Concerns regarding the organ trade were the primary motivation behind the International Congress on Ethics, Justice and Commerce in Transplantation, held in Ottawa in 1989 to debate the ethical, legal, cultural and religious aspects of transplantation. The congress attracted contributions from a wide range of experts, as well as leaders from the main religions (Dossetor et al. 1990). It showed the difficulty of reading consensus on valid ethical principles across different cultural traditions (Dossetor and Manickavel 1992). Pressure from scientific societies (Istanbul Declaration 2008) has subsequently resulted in resolutions by international political organizations to prevent and fight organ trafficking (Council of Europe 2018, United Nations 2019).

References cited

Ad Hoc Committee. 1968. A definition of irreversible coma. Report of the Ad Hoc Committee of the Harvard Medical School to examine the definition of brain death. JAMA 205: 337–340.

Allepuz, A., M. Espallargues and O. Martinez. 2009. Criterios para priorizar a pacientes en lista de espera para procedimientos quirúrgicos en el Sistema Nacional de Salud. Rev Cal Asist. 24: 185–191.

Aristotle. 1934. Nichomaquean Ethics. Harvard University Press, London.

Arredondo, A., R. Rangel and E. Icaza. 1998. Costos de intervenciones para pacientes con insuficiencia renal crónica. Revista de Saúde Pública 32: 556–565.

Bazyar, J., M. Farrokhi, A. Salari and H.R. Khankeh. 2020. The principles of triage in emergencies and disaters: a systematic review. Prehosp Disaster Med. 35: 305–13.

Benach, J. and J. Alonso. 1995. El plan de salud del Estado de Oregón para el acceso a los servicios sanitarios: contexto, elaboración y características. Gac Sanit. 9: 117–125.

Bentham, J. 1987. An Introduction to the Principles of Morals and Legislation. Penguin Books, New York.

Buchanan, J. 1986. Liberty, Market and State. Harverster Press, Brighton.

Calsamiglia, X. 1996. Equity criteria and the setting of priorities for health policies. pp. 13–41. *In*: Abel, F. (ed.). Allocation of Resources and Choices in Health Care. Institut Borja de Bioètica y Fundación Mapfre Medicina, Barcelona, Spain.

Churchill, L.R. 1992. Theories of Justice. pp. 21–34. *In*: Kjellstrand, C.M. and J.B. Dossetor (eds.). Ethical Problems in Dialysis and Transplantation. Kluwer, Dordrecht, Netherlands.

Council of Europe. 2018. Guide for the Implementation of the Principle of Prohibition of Financial Gain with Respect to the Human Body and Its Parts from Living or Deceased Donors. Council of Europe, Strasbourg, France.

Declaration of Istanbul on organ trafficking and transplant tourism. 2008. Transplantation. 86: 1013–1018.

Dossetor, J.B., A. Monaco, P. Stiller and R. Calvin (eds.). 1990. Ethics, laws and commerce in transplantation. a global view. Transplant Proc. 23: 4–123.

Dossetor, J.B. and V. Manickavel. 1992. Commercialization: The buying and selling ok Kidneys. pp. 61–71. *In*: Kjellstrand, C.M. and J.B. Dossetor (eds.). Ethical Problems in Dialysis and Transplantation. Kluwer, Dordrecht, Netherlands.

Dworkin, R. 1981. What is Equality? Part 2: Equality of resources. Philosophy and Public Affairs 10: 283–345.

Edwards, R.T. 1999. Points for pain: waiting list priority scoring systems. BMJ 318: 412–414.

Emanuel, E.J., G. Persad, R. Upshur, B. Thome, M. Parker, A. Glickman et al. 2020. Fair allocation of scarce medical resources in the time of Covid-19. N Engl. J. Med. 382: 2049–2055.

Engelhardt, T.H. 1986. The Foundations of Bioethics. Oxford University Press, New York.

Espallargues, M. 2018. Priorización de intervenciones sanitarias: de la teoría a la práctica. pp. 44–68. *In*: Puyol, A. and A. Segura (eds.). Prioridades y políticas sanitarias. Cuadernos de la Fundación Víctor Grifols i Lucas nº 48, Barcelona, Spain.

Fabre, J., P. Murphy and R. Matesanz. 2010. Presumed consent: a distraction in the quest for increasing rates of organ donation. BMJ 341: c4973.

García Lizana, F., L. Santana Cabrera, J.C. Martín González and M. Sánchez-Palacios. 2010. Limitación del tratamiento en una unidad de cuidados intensivos. Med Clín (Barna). 135: 573–4.

González-Castro, A., J.C. Rodríguez Borregán, O. Azcune Echeverria, I. Perez Martín, M. Arbalan Carpintero, P. Escudero Acha et al. 2017. Evolución en las decisiones de

limitación de los tratamientos de soporte vital en una unidad de cuidados intensivos durante una década (2005–2014). Rev. Esp. Med. Legal 43: 92–8.
Gracia, D. 1989. Fundamentos de Bioética. Eudema, Madrid.
Gracia, D. 1996. Priorities in Health Care. pp. 201–205. *In*: Abel, F. (ed.). Allocation of Resources and Choices in Health Care. Institut Borja de Bioètica y Fundación Mapfre Medicina, Barcelona, Spain.
Hadorn, D.C. 2000. Setting priorities for waiting lists: defining our terms. CMAJ 163: 857–860.
Honas, H. 1984. The Imperative of Responsibility. In Search of an Ethics for the Technological Age. University of Chicago Press, Chicago.
Ingelfinger, J.R. 2005. Risks and benefits to the living donor. N Eng. J. Med. 353: 447–449.
Kant, I. 2005. Fundamentación de la metafísica de las costumbres. Editorial Tecnos, Madrid.
Kaplan, R.M. 1993. Application of a general health policy model in the American health care crisis. Journal of the Royal Society of Medicine 86: 277–281.
MacCormick, A.D., W.G. Collecut and B.R. Parry, 2003. Prioritizing patients for elective surgery: a systematic review. ANZ J. Surg. 73: 633–642.
MacIntyre, A. 1981. After Virtue. Notre Dame University Press, Indiana.
Martin-Fumadó, C., E.L. Gómez-Duran and M. Morlans Molina. 2020. Consideraciones éticas y médico-legales sobre la limitación de recursos y decisiones clínicas en la pandemia del COVID-19. Rev. Esp. Med. Legal. 46: 119–26.
Miller, F.C., M. Nair-Collins and R.D. Truog. 2021. It is time to abandon the dogma that brain death is biological death. The Hasting Center Report 51: 18–21.
Mollaret, P. and M. Goulon. 1959. Le coma dépassé. Rev. Neurol. (Paris) 101: 3–15.
Morlans, M. 1996. La responsabilité sociale du médecin: une constriction de son rôle fiduciaire envers le malade? pp. 141–155. *In*: Abel, F. (ed.). Allocation of Resources and Choices in Health Care. Institut Borja de Bioètica y Fundación Mapfre Medicina, Barcelona, Spain.
Morlans, M. 2011. La donación de órganos y los trasplantes. pp. 461–485. *In*: Boladeras, M. (ed.). Bioética: la toma de decisiones. Ed. Proteus, Barcelona, Spain.
Morlans, M. 2019a. El trasplante de donante vivo: un reto clínico y moral. Boletín de la Academia Chilena de Medicina 56: 97–106.
Morlans, M. 2019b. Experiencia nacional de trasplantes en España. Boletín de la Academia Chilena de Medicina 56: 201–208.
Morlans, M. 2019c. Fundamentos éticos de los trasplantes. Boletín de la Academia Chilena de Medicina 56: 155–163.
Mill, J.S. 1982. On Liberty. Penguin Books, New York.
Mill, J.S. 1987. Utilitarianism. Penguin Books, New York.
National Commission. 1978. Principios éticos y orientaciones para la protección de sujetos humanos en la experimentación. DHEW Publication No. (OS) 78-0012.
NHSBT. 2019. Kidney Transplantation: Deceased Donor Organ Allocation. Policy Pol 186/10. https://nhsbtdbe.blob.core.windows.net/umbraco-assets-corp/16915/kidney-allocation-policy-pol186.pdf.
Nozick, R. 1974. Anarchy, State and Utopia. Basic Books, New York.
Nussbaum, M. 1992. Human functioning and social justice. Political Theory 20: 202–246.
OPTN. 2022. Policies. Organ Procurement and Transplantation Network. https://optn.transplant.hrsa.gov/media/eavh5bf3/optn_policies.pdf.
Puyol, A. 1999. Justícia i Salut: Ètica per al racionament dels recursos sanitaris. Universitat Autònoma de Barcelona, Bellaterra.
Puyol, A. 2001. El discurso de la igualdad. Crítica, Barcelona.
Rawls, J. 1971. A Theory of Justice. Harvard University Press, Cambridge.
Royal Decree 2070. 1999. Por el que se regulan las actividades de obtención y utilización clínica de órganos humanos y la coordinación territorial en materia de donación y trasplante de órganos y tejidos. Boletín Oficial del Estado 3: 179–180.

Richards, J.R. 1992. From him that hath not. pp. 53–60. *In*: Kjellstrand, C.M. and J.B. Dossetor (eds.). Ethical Problems in Dialysis and Transplantation. Kluwer, Dordrecht, Netherlands.

Rudge, C.J. 2018. Organ donation: opting in or opting out? Br Gen Pract. 68: 62–63.

Sampietro, L. and M. Espallargues. 2004. Opiniones, vivencias y percepciones de los ciudadanos en torno a las listas de espera para cirugía electiva de catarata y artroplastia de cadera y rodilla. Aten Prim. 33: 86–94.

Sandel, M.J. 2020. The Tyranny of Merit: What's become of the Common Good? Penguin Random House, London.

Sen, A. 2009. The Idea of Justice. Penguin Books, New York.

Simón, P. 2000. El consentimiento informado. Historia, teoría y práctica. Triacastela, Madrid.

Solans-Domenech, M., P. Adam, C. Tebé and M. Espallargues. 2013. Developing a universal tool for the priorization of patients waiting for elective surgery. Health Policy 113: 118–126.

Strathern, M. 2011. Human Bodies: Donation for Medicine and Research. Nuffield Council on Bioethics, London.

Tebé, C., M. Comas, P. Adam, M. Solans-Domenech, A. Allepuz and M. Espallargues. 2015. Impact of a prioritary system on patients in waiting lists for knee arthroplasty. J. Eval. Clin. Pract. 21: 91–96.

The Multi-Society Task Force on PVS. 1994. Medical aspects of the persistent vegetative state. N Engl. J. Med. 330: 1499–1508.

Truog, R.D. 2005. The ethics of organ donation in living donors. N Eng. J. Med. 353: 444–446.

United Nations. General Assembly. 2019. Strengthening and promoting effective measures and international cooperation on organ donation and transplantation to prevent and combat trafficking in persons for the purpose of organ removal and trafficking in human organs A/RES/73/189.

Valls, E. 2009. Història del trasplantament d'òrgans a Catalunya. Edicions 62, Barcelona.

Williams, A. 1985. Economics of coronary artery by-pass grafting. BMJ 291: 386–389.

Willis, B.H. and M. Quigley. 2014. Opt-out organ donation: on evidence and public policy. J. R Soc. Med. 107: 56–60.

Chapter 7
Social Bioethics

Begoña Román Maestre

Introduction

This article has two main objectives. The first is to explain what social bioethics means. The second is to explain whether the adjective "social" does nothing more than complement clinical bioethics or whether it deserves its own field. As a mere complement, "social" is a dimension of the bio-psycho-social model. However, if social Bioethics has its own scope, it must be distinguished from clinical bioethics and must be understood in a broader sense that affects the very concept of health. In this case, this social Bioethics has its own criteria beyond the classical principles of Medical Ethics. In the first narrow sense, the future of bioethics will depend on the integration of health and social services. However, in a broader sense, social Bioethics encompasses more topics, for example, animal ethics or environmental ethics.

This document consists of four parts. The first part addresses the question of what social bioethics means. It is known that 80% of the determinants of health are social (WHO 2022). Poverty, addictions, violence in its different forms, are chronic social problems that cause diseases, illnesses and ailments. For this reason, it is important that health professionals tackle the social perspective and that social and health policies work together.

The second part deals with the important role that social professionals have in the design of health Policies and Services. The approaches of social professionals give a level of realism about the daily circumstances of people. This is especially important in preventive health campaigns,

Universitat de Barcelona, Spain.
Email: broman@ub.edu

mental health services and public health. Mental illness and loneliness in the elderly are two useful cases to illustrate how essential this social Bioethics is.

The third part deals with the specific criteria that are relevant in social bioethics: vulnerability, care, empowerment and social justice and recognition. Social Bioethics can contribute to achieving more levels of Health with better levels of social justice and care.

The fourth part deals with a broader sense of social Bioethics that includes Animal Ethics and Environmental Ethics. Our conceptions of animals and the environment determine our relationship with them. And in turn, our relationship with them depends on the following: if we think they deserve respect or are merely a means for our interest or benefit.

What social Bioethics means

The answer to the question of what social Bioethics means depends on the meaning and scope of the word "social". As an adjective, "social" can be understood simply as a complement to clinical bioethics. From this point of view, as a mere complement, the "social" is a dimension of the biopsychosocial model (Engel 1977). In this sense, the objective of social Bioethics is to broaden the view and observe the social environment where the patient lives. However, in this narrow meaning of social, as one more dimension among three, the leading role is played by the biological ones. It is very common to pay attention to the social aspect provided that it can improve or harm health. But the focus is on the biological dimension. The social dimension is just a stage, it is important, but not the most important.

Doctors look beyond the biological or psychological dimension because people's lives are situated in the social environment (financial and political) and this has a great influence on health and its recovery. In this sense, the disease, diagnosis or treatment may be the same, but the social context in which the patient finds himself and his emotional and psychological experience will be different. Knowing all this information about the social dimension is crucial for treatment, recovery or care.

It is known that 80% of health determinants are social (WHO 2022). However, most of the time this attention to the social dimension is too individualistic, case by case. Social determinants of health are the conditions in which people are born, live, work and age. They include different factors, for instance, environment, socioeconomic status, neighborhood, education, employment, social support networks or access to health care. All of them depend on social structures, not only on the personal will.

Better coordination between social and health services is needed and at this moment they work very separately. When the visions of health professionals include social aspects, and social professionals work together in the health system, the results in health and wellbeing terms are better. Nevertheless, both of them often work together only with an individual patient.

When "social" is more than an adjective and it means the context where every action has place, another is the meaning; social Bioethics takes its own realm and it must be distinguished from mere clinical Bioethics. The development of community health must be complemented with this social Bioethics. In this case it is not a matter of widening our perspective, but opening our minds.

Despite the fact that human circumstances are always social, and these intervene in people's health, people, in turn, decide on their lifestyle and how much to appreciate their health. Moreover, the same concept of health depends on social conceptions. For instance, in many societies preventive medicine is not an important part of the public health system due to a lack of money. For other people, and for different reasons, preventive medicine can even create hypochondria among people and more expenses than real benefits in terms of health or well-being. The importance of the social dimension in health is so big that its management should not be case by case, as is usual in clinical Bioethics, but from a communitarian and public health approach. Nevertheless, the development of these approaches is lower than the case by case, face to face medicine. In turn and logically, this individualistic approach is due to social assumptions. The individualism of our current societies explains that the principle of autonomy of biomedical ethics has been for many years the prevalent one.

For many years, social aspects used to focus on vulnerable people that are poor people or people at risk of social exclusion due to stigmatized illness such as mental illness or disabilities. For many centuries, these people were not interesting for medicine nor policy, but for charities, religious centres and non-government organizations. With the development of social policies, social professions such as social workers or educators emerge independently of health services. The social aspect becomes more and more important when social justice demands care, and not only from nurses or technical care, but care regarding the relationship, inclusion and so on. In short, social Bioethics, in a wider sense, become at stake when the same society with its structures causes illness and suffering.

When the visions of health professionals include social aspects and social professionals working together in the health system, the results in terms of health and well-being are better. However, at the moment they work very separately, so better coordination between health and social services is needed. Also, when the two do sometimes work together, it only happens with an individual patient from an individualistic approach.

The development of community health must be complemented with social bioethics.

Social Bioethics can contribute to achieving more levels of Health with better levels of social justice and care. In turn, these levels depend on the attention to the social basic structure and how vulnerability, empowerment and recognition are managed within it.

The role of social professionals in the design of health Policies and Services

There is no doubt that the biopsychosocial model and patient-centred care have contributed to improving patient care by the incorporation of social and psychological aspects. People's lives pass by as a unit, as a life story where it is not easy to distinguish what the real cause of the illness or the problem of health is. The notion of "quality of life" contains objective ingredients (it is difficult to feel well with a fever of 40 degrees Celsius, or with hunger from scarcity of food); but another ingredient is a subjective one. Suffering is not pain (Ricoeur 2019), only the person who suffers can say whether the level of his or her suffering is unbearable or not.

The incorporation of psychologists and social workers or social educators into the national health system in order to carry out socio-educational intervention or mediation, if necessary, is good news. This integrative attention is crucial for obtaining high levels of patient quality. Social professionals' approaches, paying attention to the daily circumstances of the people, can avoid not only illness, but promote social cohesion and integration. This is especially important in preventive health campaigns, mental health services and public health.

Mental illness and loneliness in old people are two cases useful to illustrate how essential this social Bioethics is. In the case of mental illness, if biological treatment is excellent, but it is not accompanied with social inclusion (needing social policies and social professionals) the results will be unsatisfactory and inefficient for all affected people (the patient and her relatives, professionals and organizations). In the case of loneliness in old people, they are sometimes being treated as ill people when often the causes of their lack of wellbeing are social.

In spite of that, this model follows a micro level patient-centred and, in several cases, community-centred attention. Moreover, care is too focused on the biological dimension of health and from evidence-based medicine. The scope of social glance must be wider than this biopsychosocial model, because "social" is not only a stage or mere circumstances. These circumstances are part of us, and if we do not save them, we do not save ourselves (Ortega y Gasset 2014).

Contemporary societies have often turned the condition of dependence, fruit of our vulnerable condition, into a shameful condition.

Mirrored by this excessive self-made independence and self-sufficiency we become more vulnerable. If it is a shame to be dependent and to need each other, it is very difficult to forge solid bonds and trust. This has been seen by several criticisms of the biomedical principle of autonomy: too much liberal individualism, excess of rational competence. Descartes' mistake was not only to reduce the individual to be a *res cogitans*, but to be someone who is self-sufficient and independent at all. This liberal anthropological conception was radicalized by the Sartrean subject, son of the Nietzschean *Übermensch*.

This anthropological and social paradigm is false (the evolution of the human species proves it) and ethically wrong, by hindering the pursuit of a good and fair life. That is why we need to replace the metaphor of the cowboy in an endangered natural world with that other one, the metaphor of the astronaut, who lives in a technical world and has to work as a team. In fact, the cowboy does not need anyone, he always finds land, food, he is young, with a great ability to withstand adversity alone on a lush planet where he always has at his disposal everything he needs for living (Rosa 2021). This fiction of possessive individualism fails anthropologically and socially. The astronaut depends on the ship, the care of this environment and the relationships with the other members with whom he lives, where each has a role to play and where good understanding between them is essential for all of them. The goal is the good life not mere survival. Therefore, as we will see later, social Bioethics is open to contemplating our relationships with other non-human animals and with the environment.

This individualistic paradigm, in technical and bureaucratized societies, creates contracts, but is not able to generate communities of belonging. That is why the result is greater personal and social vulnerability. Following Ortega y Gasset (2014), when the environment and the circumstances fail, so does the self. The most we achieve is to create defensive communities that unite reactively, against someone, without much shared horizons of a world for all (Sennett 2000).

When this vulnerability is not just of the individual, but of the communities in which he or she inevitably lives, we speak of social vulnerability. Since the person is in-divided in relation, we can neither divide it nor in parts or dimensions (bio-psycho-social-spiritual, with demarcated, separated scripts), nor extract or abstract from its relations with others. Individuals are social micro institutions, we are effects of our history, and the concepts of health or illness (the consideration of homosexuality as a disease, for instance), can show how medicine is influenced by social mentality or political ideologies. The history of medicine, nursing or psychology is full of this influence.

In order to defend how important the vision of social Bioethics as a critical and wider approach is, two topics will be dealt with: physical and mental disability and the subject of sex and gender.

Regarding physical and mental disability, it is important to remind oneself of the three models of how disability in their different forms has been dealt with through time. The predominant model has been the biomedical one. For this model, disability is an illness and deficiency that must be treated, if it is possible, with medical treatments or surgery. From this model, more research is needed to improve the life of disabled people and, as soon as possible, to overcome disability that is considered as something wrong. The lives of these people take place in closed institutions, in order to prevent risks or dangers.

The second model is a reaction against this medical model. For this second model, the social-hermeneutic one, disability is not an illness. The main problem is the rejection from a majority of society that does not accept human diversity. Changes must happen in society not in the different body, mind or level of capacities or skills of disabled people.

The third model is based on rights. In a synthesis of the other former models, for the latter, disabled people deserve a special attention for their disability, but in order to be able to exercise their rights as an independent living. They need support and not just paternalistic and total protection. Disabled people deserve their independence without the legal incapacity that impedes a personal and appropriate life. This model of rights is in the spirit of the United Nations Convention on the Rights of Persons with Disabilities (United Nations 2006).

There is a debate about the appropriate name: impairment or functional diversity (Canimas 2015). If we think that an impairment should be interpreted only as diversity, what arguments would we have against parental decisions not to correct a physical, intellectual or developmental impairment in their children through an effective and safe medical/surgical intervention? What could be the arguments to avoid those parents causing their children impairment? This is the case of a deaf same-sex couple who, through assisted reproductive technology, were able to choose that their son was deaf like them (Sandel 2009). For them deafness is a cultural identity, not an illness at all.

Another problem will be the justification of the need for affirmative action such as more resources, attention, medical and technological research for people with functional diversity. And last but not least this debate about names (functional diversity or impairment) doesn't find consensus among the affected people. Some of them fear the passing from an overprotection policy to an under-protection one until reaching levels of abandonment. It can happen, for instance, if following the model of rights, that the Convention of United Nations (2006) is implemented by law without any social changes and financial resources for independent life. This could be the case of impossibility of involuntary admission to hospital for people with mental illness in a moment of acute crisis (Comitè de Bioètica de Catalunya i Comitè d'Ètica dels Serveis Socials de Catalunya 2021).

Issues regarding sex and gender have become more and more relevant in the last 20 years. There is no doubt that feminist movements and queer philosophy (Butler 1999) have helped. Sex and gender are not equivalent. By sex we understand the biological distinction of an individual, generally based on organic (genital) and chromosomal characteristics. Gender is the set of characteristics of cultural origin related to patterns of behaviour and identity from which the distinction between men, women or others is socially established. As it is a sociocultural construction, gender is a dynamic concept that varies over time.

Although socially the male gender is attributed to people with the male sex, and the female gender to those of the female sex, other identities have appeared, such as "queer, genderqueer", which question both the gender binarism and the medicalization due to gender migration and the reproduction of gender roles. They claim that people can exhibit traits of the two characteristics marked, until now, as exclusive to a certain gender.

From a social bioethical approach, it is not necessary to go into the reasons why people want to be trans, which can be very complex, varied and personal. Nor is it about focusing on it as a biological problem, since biological data does not directly define the gender identity that one wants. Rather, the important question is that people who request help in the search for an identity different from the one they were born with should be treated without being stigmatized. Nor should it be immediately related to a mental problem or diagnosed as gender dysphoria. This clinical category could be a part of the problem more than a solution and it can add more suffering. Trans people do not necessarily ask for care because they have a disease or a disorder, but in most cases it is because of the suffering caused by the obstacles they encounter in the free development of their most fundamental rights, as well as in access to true information about treatments or services.

The process of defining the sexual identity of transgender is a difficult personal path to which is added significant social pressure, stigmatization, marginalization and, very often, violence. This can lead to conflicts and symptoms that need to be treated. Anxiety, stress, depression, self-harm and suicide are more frequent in the trans community in which, in addition, intersectionality (ethnicity, religion, culture, poverty, etc.) can complicate and aggravate the situation of these groups, increasing their vulnerability (Turban et al. 2022).

Since human sexuality is a great drive that manifests in behaviours that, like many others, can lead to difficulties and need accompaniment, trans people, like others, can also have internal obstacles in the experience of their sexuality. However, the external obstacles are more important from a social Bioethics' perspective. The objective of the services and the professionals who attend people with discomfort with their sexual or gender identity is to accompany them in these processes if they want to change to another gender and to manage the level of changes in their bodies (James et al. 2016).

If people diagnosed with gender dysphoria (there is a real debate about this diagnosis as a mental illness) are adults, the mere reference to the patient's autonomy is enough to respect the personal decisions of this competent adult. The question is more complex in the case of minors. Suffering due to sex and gender usually begins with puberty. There is no agreement about whether or not it is convenient to give hormonal treatments to minors. The Nuffield Council (2022) has stopped its research regarding trans minors while it waits for more evidence. Depending on social pressure, ideologies and governments, the accompanying of trans minor is very variable in every culture.

Several relevant criteria in social Bioethics

Social Bioethics searches to contribute to achieve more levels of Health with better levels of social justice and care. In this sense, as it is logical, the most relevant criteria in social Bioethics, but not limited to, are vulnerability, care, empowerment and social justice and recognition.

Vulnerability

Vulnerability means capacity of harm or to be harmed. Vulnerable is not only the victim, who suffers an evil, a pain, but also the perpetrator, because he or she also harms himself or herself even in his or her moral value. But the damage, the wound, the humiliation is not only bad for the evil or bad feelings it occasionally generates, but for those who will come as a reaction, in the chain of resentments impeding social cohesion. Despite this universal condition inherent in human beings, vulnerability is not the same for everyone, it depends on personal and social resources. Adversity does not affect everyone equally and it tests defences and resources to deal with it. We can tell in pain and suffering our condition of finitude, our vulnerability (Han 2021).

In social Bioethics we don't talk about individual or personal vulnerability but a social one. In this sense, a lot of people have bad luck. Depending on the place of birth, their family, their cultural level and so on, their future is closed, like a destiny. Vulnerable groups deserve positive discrimination. And all actions with and for them must avoid paternalism.

Social Bioethics tries to prevent more inconvenience for disadvantaged people. For instance, promoting access to culture or concerts requires destroying physical barriers for disabled people, or lowering prices for people from lower-income backgrounds. But this is insufficient, because access to culture is needed for an education. Accessibility for vulnerable people is not only a matter of barriers or prices; it's a matter of readiness.

The moral development of a society is shown in how it takes care of the vulnerable people in it. But the forms of vulnerability are new: sometimes they are not only due to health reasons, but also to access to work, housing and social inclusion. Sometimes the vulnerability is individual, other times it is of the group, be it collective or ethnic group; but in others it is national. The human condition is interdependent. In a globalized world, we have seen it with COVID-19 and we see it with the current war in Ukraine. Social justice is not only at the country level, but has to broaden its horizons to global and even intergenerational justice for future generations.

Being vulnerable means that we need care and affection because we are affected by many things, the main one being the lack of affection. A society that shakes human bonds, that dismantles the community (Bauman 2005), makes it really complicated to cooperatively solve its problems. That is why vulnerability requires care, which means responding to the fragility of the other person.

Care

Excessive stress from pain or suffering; for fear of poverty or becoming a burden to others are the main causes of vulnerability. Social policies can prevent much of that suffering. Care seeks stability, calm, capabilities and bonds (Román 2016). Caring needs more than feelings of empathy or compassion. It needs knowledge and organization. This requires caregivers to have skills, knowledge and confidence. Care has a political dimension (Tronto 1993). Care and justice must go together. It is not a good idea or strategy to separate them.

Care is not a matter of feminism or feelings. Gilligan's (2016) different voice is not the voice of women, it was the voice of humanity. And social policies are not just a matter of a good portfolio of services in the home. Care means relationship. Care takes place in close, rooted communities, with knowledge of their territory and cultures within it. People live their lives in a social environment beyond their own health and home. Reconciling the big with the small, domestic and common, forces us to think better about responsibility, because responsibility is the duty to take care of vulnerability.

Social Bioethics implies thinking about our communities. The community dismantled by the macro and by the mega in our globalized world forces us to rethink the traditional political categories of territory, state and nation. The COVID 19 pandemic must be a lesson to learn about cooperation and interdependence. It is not a question of closing in on oneself or in one's own country. We must trust humanity, not from impersonal ties but from fraternal and cosmopolitan ones. We all need different kinds of communities of belonging. We need to create a patchwork blanket in which each different piece is known the same and is linked with others from other territories and customs but all of them share the status of citizens of the world. This change of mentality is achieved not only with ideas, but also with experiences of proximity and hospitality. Care is the word that sums it up.

Care implies trust not only by people to people. We must believe in our organizations and in our organized agency. The 21st century started with a great crisis in which trust is a fundamental key to get out of it. But today's trust must be that of a knowledgeable society, it cannot be a blind faith typical of the Middle Ages. That trust must be based on the fact that we are free, we have *place* to manage *and* change things, and we are responsible for what we do or we fail to do. In addition, it must be a trust based on scientific knowledge and transparency about processes.

Empowerment

Personal vulnerability is not just about caring for dependency, it is about promoting autonomy. Some social policies have carried out paternalistic approaches on fragility and dependency, instead of promoting or maintaining degrees of autonomy and competence. These approaches focused so much on the shortcomings that they did not address the emancipation and empowerment of the individuals, groups and communities with whom the professionals work. It is not so much about showing their flaws or shortcomings, what they cannot do, but about promoting what they know and can do. It is a matter of fostering capabilities.

People must be able to develop their potential, their capabilities to carry out the life project they have decided on. And all of this does not depend only on the individual will. The social context can promote or block it. Autonomy is not only, as Kant (2000) conceived it, a "sapere aude", but it also means being able to do, to become real, what one wants and can do. The capabilities approach as conceived by A. Sen and M. Nussbaum helps materialize a concept of autonomy with a material foundation and rooted in multicultural societies (Nussbaum 2007).

Social justice and recognition

The incidence of luck in a people is inversely proportional to the level of justice of the society in which they live: it is "bad luck" to be disabled, to suffer from an illness that condemns one to dependency, but more unfair than this chance would be to condemn one to undignified treatment. Quoting J. Rawls (1999) and his veil of ignorance, everyone would choose, if they did not know their biological-social lottery, to live in a fair and just society in solidarity rather than in an unfair society. In the former, those most advantaged by the biological-social lottery have to help reduce the disadvantages of those least advantaged by such a lottery (Rawls 1999); in an unfair society the law of the jungle reigns, and in it makes no sense to speak of ethics or human dignity; in it, the dependents or the disabled, for example, are excluded by simple "natural selection".

It is necessary to demand the distribution of limited resources, fairness, equality, impartiality. But justice must be distinguished from the notion of good living and quality of life, precisely to make possible the peaceful coexistence of plural morals, differentiated in substantive conceptions of the good, in today's societies. Justice allows the pursuit of the good life without specifying, only limiting, what it consists of. Social justice implies solidarity and recognition.

Social cohesion only triumphs where the other is one of us. For that, dialogue must be better institutionalized. Beyond the majority vote, it is necessary to avoid majority preferences that may contradict the rights of all. Sometimes the majorities themselves are not aware that they are voting against their own interests due to a change of chance. Solidarity makes the lack of justice explicit and reminds us of the need to help one of us. Humanism is characterized precisely by considering every human as being sacred, whether a brother for being a son of God, as in the Christian tradition, or, as in utopian socialism, for being a brother or sister of common humanity.

Recognition is always a struggle (Honneth 1997) to be truly treated as equals, a hope for justice for overcoming historical discrimination. Human beings project and create the future. Precisely because it is a question of people's past not being their destiny, the struggle for recognition implies hope.

This principle is especially important in vulnerable populations: it is about fighting bad luck from justice, with the hope that there is an alternative to the contemporary condition. This is contrary to Margaret Thatcher's famous "There is no alternative" (TINA), which in times of crisis is heard again in the voices of politicians. We must insist that ethics and politics are in the realm of contingency, of what depends on us, not of necessity.

Vulnerability in the 21st century is presented in the form of excessive and continuous provisional, disempowered, lack of protective ties. Alleviating vulnerability requires creating relatively stable environments, empowering subjects to manage change, and building bonds of trust in a networked community that never leaves anyone exposed to terrible and inhumane outdoor.

All of these principles are *prima facie*. Its primacy will depend on the cases and circumstances. The virtue of prudence plays a very important role in this. Prudence is like a strategic intelligence. It is a moral duty to be intelligent and not fall into the naivety that good will is enough. Efficiency is also essential in applied ethics (Bayertz 2003).

The criteria of vulnerability, empowerment, care, social justice and recognition help to face ever-changing scenarios and forms of personal and social fragility. The ethical and political will to put the brain and the heart at the service of a better human world will continue to be fundamental.

A broader sense of social Bioethics: Animal Ethics and Environment Ethics

A broader sense of social Bioethics could include topics such as Animal Ethics and Environmental Ethics. In fact, our conceptions of animals and the environment, which depend on social values, determine our relationship with them. And in turn, our relationship with them depends on that: if we think they deserve respect or they are mere means for our interest or benefit.

Animal ethics focuses on the right and wrong way to treat animals. Several theories defend that animals have rights and this is the main motive to avoid any path to animal abuse or cruelty. For other theories, the important thing is not if animals have or do not have rights, but if all of them are sentient, and this is the main point to recognize that they have moral dignity. Dignity is not an exclusive attribute of human beings (rational and superior). Dignity depends on being sensitive. Their moral worth or rights is independent of their utility for humans (Regan 2016).

However, animal ethics doesn't defend a negative approach, such as not doing them harm or preventing them from suffering. Animal ethics proposes that their most basic interests should be afforded the same consideration as similar interests of human beings (Singer 1990). Animal ethics has expanded the ethical issues by criticizing the ever and ever unquestionable superiority of human beings.

In this sense, several theories criticize the use of animals in the circus, or the life in the zoo and other shows. Animal lovers are criticized too in their ownership of animals. Choosing to have pets, choosing what kind of pets to have and choosing what kind of life for them are crucial moral concerns.

This animal ethics needs a policy (Donaldson and Kymlicka 2011). We must share our spaces with them and more and more humans are arrogating all territories and as a consequence, jeopardize wildlife and condemning it to extinction. Animal ethics not only denounces cruelty in the treatment of animals in big farms, but also disputes animals as human food.

As we can see, all these topics are related to the social conceptions of animals; and these conceptions are implemented in habits, traditions, etc. The goal is to urgently remove them because they cause animal suffering.

Environmental ethics studies the conceptual foundations of environmental values as well as more concrete issues surrounding actions and policies to protect and sustain biodiversity and ecological systems. Environmental ethics have been developed from two very different perspectives: anthropocentrism and biocentrism. The former is human-centred; the latter is nature-centred. Depending on what perspective is adopted, the human-nature relationships change as well as the priorities and the policies.

For anthropocentrism we must unconditionally take care of the environment because it is our home, and human life on Earth must be preserved. In this approach the environment value is not an intrinsic one. The absolute value is the human being, and only because the environment is necessary for human life, we must respect the environment. Moreover, the priorities change depending on the sacrifices that human beings must make in order to preserve other life on Earth and the quality of life of the next generations of humans. For anthropocentrism sacrifices in quality of life in the present must be required in order to guarantee generational justice. In other words, the future generations of humans have the right to live on Earth, to enjoy their time on Earth. This opportunity must have good possibilities for searching for the good life. We are responsible to bequeath to future generations a healthy, habitable and beautiful world (Jonas 1995). This is a duty for all of us, as humanity, not only for every individual.

For the biocentric approach, nature has an intrinsic value, and for this motive we must preserve, for instance the wilderness and, if it were necessary, to reduce the presence and impact of humans on Earth. In this approach the movement of deep ecology is remarkable. This promotes the inherent worth of all living beings regardless of their instrumental utility to human needs, and the restructuring of modern human societies in accordance with such ideas (Naes 1973).

Our conceptions of animals and the environment determine our relationship with them. And in turn, our relationship with them depends on social beliefs, attitudes and mentalities. Animal ethics and environmental ethics ask for changes in human lifestyles and from their approaches, societies must be changed and penalized if they don't contribute to preserving the possibility of life, both human and nonhuman beings.

The One Health concept summarizes the approach that recognizes that the health of people is closely related to the health of animals and our shared environment. But it focuses on zoonotic diseases or safe food, etc. Instead, social Bioethics involves critical thinking about society, health, and lifestyles. The main point is not health, but the reasons for living and the kind of world we want to live in and maintain. Health is an important ingredient of these reasons to live. But it is each one who must decide what level of importance health has in his life. With one limitation: not to harm others, as recalled by the Hippocratic Oath in 5th century B.C. But today "others" means more affected beings. We are in charge of our world, its present and future. This broader perspective is that of social Bioethics.

References cited

Bauman, Z. 2005. Amor líquido. Acerca de la fragilidad de los vínculos humanos. Fondo de Cultura Económica, Madrid.

Bayertz, K. 2003. La moral como construcción. Una autoreflexión sobre la ética aplicada. pp. 47–70. *In*: Cortina, A. and D. García-Marzá (eds.). Razón pública y éticas aplicadas: los caminos de la razón práctica en una sociedad pluralista. Tecnos, Madrid, Spain.

Butler, J. 1999. Gender Trouble. Feminism and the Subversion of Identity. Routledge, New York.

Canimas, J. 2015. ¿Discapacidad o diversidad funcional? Siglo Cero. 46: 79–97.

Comitè de Bioètica de Catalunya i Comitè d'Ètica dels Serveis Socials de Catalunya. 2021. Sobre les hospitalitzacions involuntàries en l'àmbit de la salut mental. https://canalsalut.gencat.cat/ca/detalls/noticia/Sobre-les-hospitalitzacions-involuntaries-en-lambit-de-la-salut-mental.

Donaldson, S. and W. Kymlicka. 2011. Zoopolis: A Political Theory of Animal Rights. Oxford University Press, Oxford-New York.

Engel, G.L. 1977. The Need for a New Medical Model. A challenge for biomedicine. Science, New Series 196: 129–136.

Gilligan, C. 2016. *In*: A Different Voice: Psychological Theory and Women's Development. Havard University Press, Cambridge.

Han, B.Ch. 2021. La sociedad paliativa. Herder, Barcelona.

Honneth, A. 1997. La lucha por el reconocimiento. Por una gramática moral de los conflictos sociales. Crítica, Barcelona.

James, S.E., J.L. Herman, S. Rankin, M. Keisling, L. Mottet and M. Anafi. 2016. The Report of the 2015 U.S. Transgender Survey. National Center for Transgender Equality, Washington.

Jonas, H. 1995. El principio de responsabilidad. Ensayo de una ética para la civilización tecnológica. Herder, Barcelona.

Kant, I. 2000 ¿Qué es la Ilustración? pp. 25–37. *In*: Kant, I. (ed.). Filosofía de la Historia. Fondo de Cultura Económica, Madrid, Spain.

Naes, A. 1973. The shallow and the deep, long-range ecology movement: A summary. Inquiry 16: 95–100.

Nuffield Council on Bioethics. 2022. The care and treatment of children and adolescents in relation to their gender identity in the UK. https://www.nuffieldbioethics.org/

publications/the-care-and-treatment-of-children-and-adolescents-in-relation-to-their-gender-identity-in-the-uk/call-for-evidence.
Nussbaum, M. 2007. Frontiers of Justice. Disability, Nationality, Species Membership. Harvard University Press, Cambridge.
Ortega y Gasset, J. 2014. Meditaciones del Quijote y otros ensayos. Alianza editorial, Madrid.
Rawls, J. 1999. A Theory of Justice. Revised Edition. Harvard University Press, Cambridge.
Regan, T. 2016. En defensa de los derechos de los animales. Fondo de cultura económica, México City.
Ricoeur, P. 2019. El sufrimiento no es el dolor. Isegoría: Revista de filosofía moral y política 60: 93–102.
Román, B. 2016. Ética de los servicios sociales. Herder, Barcelona.
Rosa, H. 2021. Lo indisponible. Herder, Barcelona.
Sandel, M. 2009. The Case against Perfection: Ethics in the Age of Genetic Engineering. Harvard University Press, Cambridge.
Sennett, R. 2000. La corrosión del carácter. Las consecuencias personales del Trabajo en el nuevo capitalismo. Anagrama, Barcelona.
Singer, P. 1990. Animal Liberation. New York Review of Books/Random House, New York.
Tronto, J.C. 1993. Moral Boundaries. A Political Argument for an Ethic of Care. Routledge, New York.
Turban, J.L., D. King, J. Kobe, S.L. Reisner and A.S. Keuroghlian. 2022. Access to gender affirming hormones during adolescence and mental health outcomes among transgender adults. PLoS One 7: e0261039.
United Nations. 2006. Convention on Rights Convention on the Rights of Persons with Disabilities. https://www.un.org/development/desa/disabilities/convention-on-the-rights-of-persons-with-disabilities.html.
World Health Organization. 2022. About social determinants of health. https://www.who.int/social_determinants/en/.

Future Challenges

Chapter 8
Challenges in Biomedical Areas

Salvador Macip[1,2,*] and *Christopher Willmott*[1]

Introduction: an ever-changing landscape

Although people frequently picture science as some lofty and detached activity, isolated for the rest of the world, sooner or later the implications of that work permeate into society. Unless someone has had the forethought to wrestle with the social and ethical consequences of such development, then society is frequently left playing catch-up.

From early 2020, Covid19 placed health stories front and centre in the news. However, even before the pandemic, cutting-edge innovations in biomedicine would frequently feature in bulletins. Such developments, have enormous potential to improve the lives of people, but are rarely, if ever, neutral from an ethical perspective. As moral philosopher Stanley Grenz (1998:15) noted "Modern Science has placed in our hands capabilities that have aggravated long-standing ethical problems as well as introducing new quandaries". Thus, scientific advances usually bring a combination of new challenges and old dilemmas enhanced and reshaped under the light of modern times.

It is from this perspective that it is important to reflect on the current challenges in biomedical areas and try to predict which will become relevant in the near future. Given the ever-expanding breadth of modern science, this is a complex task. The UK's Nuffield Council on Bioethics,

[1] Department of Molecular and Cell Biology, University of Leicester, Leicester, UK.
[2] FoodLab, Faculty of Health Sciences, Universitat Oberta de Catalunya, Spain.
* Corresponding author: sm460@le.ac.uk

which has been a major body exploring and reporting on ethical aspects of medicine and biology for over 20 years, publishes an annual *What's on the Horizon?* document each January. The latest version includes 72 topics, split across five categories: Health & Society; Beginning of Life; Data & Technology; Research Ethics; and Animals, Food & the Environment (Nuffield 2022). This gives us a feel for the vast and diverse arenas which bioethicists needs to review and anticipate to keep up with the frenetic pace at which science is moving.

Horizon-spotting is, by its very nature, an inherently tentative activity. Even with the breadth of coverage within Nuffield's review, there is a clear possibility that other developments will emerge, apparently from nowhere, and will take centre stage. Equally, developments that looked like true game-changers will get bogged down and become irrelevant. Some will drift off into obscurity, others will need the arrival of a further innovation to come along and unlock their potential once more.

History reminds us that the consequences of an innovation may turn out to be rather darker than was originally predicted. Thalidomide, originally intended as relief for mothers suffering from morning sickness, tragically turned out to cause extensive physical disabilities for their children, prompting major reflection about the ethics and practicalities of launching novel medicines. Cane toads were introduced into Australia from South America in the 1930s with the expectation that they would be predators for beetles that were harming sugar crops. Unfortunately, the toads not only proved ineffective at challenging the beetles, but they ate different insects, jeopardising the food chain of other native species. On top of that, the toads have poisonous skin, harmful to pets and humans. In the absence of an effective predator to limit their numbers, cane toads have continued to spread across northern areas of Australia.

Could the devastating effects of thalidomide have been avoided with better designed pre-clinical and clinical trials? Could the cane toad debacle have been prevented by running better controlled pilot studies before launching a novel ecologic intervention? Aspiring to a risk-free situation is unrealistic, but this does not mean that the risks should not be kept to the absolute minimum possible with the collaboration of experts, taking into account the perspectives of scientists, ethicists, politicians and other key stakeholders.

Since it is impossible to provide an extensive review of all the bioethical issues currently being discussed in biomedical areas, this chapter will instead focus on the ethical challenges raised by some scientific advances that appear to be progressing fastest and, notwithstanding the caveat already outlined about the perils of prediction, will likely have major significance in the coming years. The survey starts with the growing emphasis of genes and genomics. Independent developments in the ability to read long stretches of DNA, and the capability to directly manipulate the genetic code within an organism, underlie many of the specific ethical questions of the present and near-future. DNA alteration will feature later in the chapter, and this will lead to consideration of potential revolutionary changes that could redefine humanity.

Challenges of the post-genomic[1] era

New tools for new times

The significance of being able to read the four-letter language of "bases" in which DNA is written, a process known as sequencing, opened a vast array of new ethical challenges in the mid-20th century. But that was just the beginning. The date 14th April 2003 may not immediately stand out, but it has the potential to be a significant landmark for humanity. On that day, the complete DNA sequence of the human genome was officially published,[2] leading commentators to suggest that this marked the birth of the "Genomic era" (Guttmacher and Collins 2003), now commonly considered the *Post*-genomic era.

DNA sequencing itself was not new in 2003; the late Fred Sanger received his second Nobel prize in 1980 for inventing a technique to 'read' the order of bases in DNA. The fundamentals of that method served the molecular biology community faithfully, up to and including the Human Genome Project (HGP). The HGP was not, of course, the first time that the connection between genes and human health, or indeed a plethora of other DNA-based insights, had been available. Elucidation of the structure of DNA by James Watson and Francis Crick, some half-century earlier, had unlocked our understanding of the molecular biology of inheritance and the connections between diseases and specific mutations in our genetic code (Watson and Crick 1953). This had been followed in the 1970s by the development of "genetic engineering" methods that made it possible to take specific sections of DNA, for example a gene, from one organism and transfer it into a different species in an intentional and controlled manner. This allowed not only for closer scrutiny of the function of such genes and the proteins they encode, but also the potential to make large quantities of medicines, such as Insulin and later the anti-malarial drug Artemesinin, in a more cost-efficient and more ethical manner.

[1] "Genome" is not a synonym for "Gene", nor is "Genomics" just another word for "Genetics". The genome is a description of the entire DNA sequence of an individual organism. It encompasses all of their genes, but also the stretches that do not appear to code for anything.

[2] Draft versions of the genome produced by the two rival teams (one publicly funded, one private) had been published simultaneously, but separately, on 16th February 2001. However, it was always known that these would require refinement, and new data is added constantly to the original database.

Genetic information could be accessed and utilised before the HGP. However, this brought a much broader sight of the entire collection of genes within the 46 chromosomes that make the human genome. All of the data examined across both of the original human genome sequencing projects[3] came from only a handful of individuals. What the HGP presented was an overview of the scale of the human genome; that it consists of approximately 3 billion "letters" in total, and features about 20,000 genes, a surprisingly small number. It also generated a reference library against which other DNA samples might be compared. To achieve this, the HGP had involved legions of scientists and data analysts working on the project for over a decade, at the cost of roughly $3 billion.

Twenty years later, however, the landscape is radically different. An explosion of ingenious alternative ways to sequence DNA have been invented. The crucial difference is that is has become genuinely feasible to read an individual's whole genome for a few hundred dollars and in a matter of hours rather than years. This opens the doorway to all sorts of applications, many of which would fall into the general category of personalised medicine, as will be discussed below.

Despite the fact that sequencing costs have fallen so dramatically, much genomic information is still derived from a slightly earlier methodology involving DNA microarrays (sometimes termed "gene chips"). These generally consist of a plate, rather like a large microscope slide, on which there are thousands of glass beads. Each bead has been pre-loaded with a short section of DNA that will match and stick to a corresponding sequence of test DNA (for example, a specific mutation) if it is present within a test sample washed across the plate. There is therefore an important distinction between a microarray-based test and a sequencing approach. In the former, a search is directed to a large number of pre-determined sites within the genome. This can be highly informative, however if there was an unexpected alteration in the sample being tested, this would not be picked up. In contrast, a sequencing-based approach allows for known and novel changes to be identified.

A revolution in medical care brings new ethical challenges

The use of both microarrays and sequencing technology has provided unprecedented access to the code of life. This can offer tremendous benefits in health care, starting with the potential for personalised medicine, the notion of tailoring somebody's treatment to their precise illness. Cancer therapy would be a prime example. Until recently, the

---

[3] The official Human Genome Project gained a rival along the way, when Craig Venter's company Celera Genomics began their own programme. The results of both were published on the same day.

capacity to categorise a cancer was restricted to identifying the organ in which the malignancy was found, and the specific type of cell within that tissue which has gone rogue. For example, the name renal cell carcinoma, implies that the cancer has emerged in epithelial cells (hence "carcinoma") lining the tubules of a kidney (hence "renal"). This would inform the treatments chosen but, in truth, the options were pretty restricted, and the side effects pretty brutal.

Now, however, genomics can provide a picture of the exact changes that are driving the cancer in a given individual, and to tailor their medicine to challenge the specific fault. It turns out, for example, that breast cancer exists in at least 10 different sub-types based on the particular main mutations that have taken place (Ali et al. 2014). Certain potential medicines would be ideal for treating some sub-types, but others may do no good, and even make the patient more unwell. Picking the right treatment for the right patient is a clear benefit of genomics. It also becomes possible to achieve much earlier diagnosis via genomics. An emerging approach involves taking a simple blood sample from a patient and looking for the minute genetic signals for a range of different cancers long before any physical changes have manifest. This type of "liquid biopsy" is especially helpful when the organ affected is harder to sample directly. Earlier diagnosis means earlier and easier treatment, and a better chance of survival.

Some of the most significant beneficiaries of a switch to a genomic approach have been individuals with so-called rare diseases and their families. Rare diseases are defined as those that affects fewer than 1 in 2000 people in the population, but may actually be much rarer than this, possibly affecting just a handful of people around the world. This makes a connection to the underlying cause very tricky. Whole genome sequencing allows comparison of the DNA of affected individual in a way that was never previously feasible. When the mystery can be solved, most rare diseases turn out to be due to single gene mutations (possibly a new mutation that was not present in either parent).

On the face of it, therefore, the shift to genomic medicine seems to be a very positive development. There are, however, a number of significant tensions at play here. Some concerns relate to the cost of genomics. Although the costs of sequencing have tumbled in recent years, it remains true that, with a finite budget, medical providers such as the UK's National Health Service (NHS) are continually having to make trade-offs between support for one area of medicine at the expense of another. Money spent on genome sequencing is money that cannot be spent on, for example, hip replacement surgery. Even time spent analysing and interpreting the data, then explaining it to the patient, is a resource-intensive process. Genetic counselling is a shortage specialism, requiring skills over and above the expertise of most family doctors. Similarly, there are questions about opportunity of access to genomic services; will this be a scenario in which the rich can access and benefit from the genomic revolution, whilst poorer individuals in the West or, especially, in Low and Middle Income Countries, will be left behind?

Alongside this, there are concerns about privacy and confidentiality. Appropriate secure handling of health information is important in general. Genetic and genomic data, however, is considered to be especially sensitive since it is inherently shared by your relatives. Discovering some genetic secret about a person may inadvertently reveal information about somebody else, possibly something such as the inevitability of their succumbing to Huntington's Disease, a fact which perhaps they were intentionally trying not to know.

There is a related concern that genomic screening may throw up something about a person that was not the initial focus of the investigation. Such 'incidental findings' may be the high probability of suffering from a devastating genetic illness about which the subject was previously unaware. It might also, however, be the accidental uncovering of a secret about the family unit. Classically, this might be the revelation that the person considered to be Daddy is not, in fact, their genetic father.

Knowing the possibility of having a genetic illness in the future might be a positive discovery, if it allows to make a lifestyle adjustment now that can help to alleviate or postpone symptoms in the future. The unexpected news is going to be particularly problematic, however, if the condition is not "actionable", in other words, that there is nothing that can be done to mitigate the anticipated suffering.

The risks of genomic data

The benefits of genomics come from an ability to share and compare sequence information. For example, Genome-Wide Association Studies (GWAS) look to uncover new information about the genetic causes of a condition by comparing the genomes of multiple people who have that condition, against those who do not. If the sufferers repeatedly have certain mutations that are not shared with the control group, this shines a spotlight on a genetic change that may be relevant to development of that illness.

The simplest associations are between single genes and a particular condition (a so-called monogenic disorder). Many illnesses, however, involve the interplay of multiple different genes and the interaction of their products with environmental factors. GWAS studies (using gene chips) are increasingly being used to develop Polygenic Risk Scores. These measures of the likelihood for someone to get a given condition are already available for some diseases, such as Type 2 diabetes, breast cancer and heart disease. Again, potential benefits include allowing someone to introduce lifestyle changes that will reduce the chances of them getting a disease or at least delaying its onset. However, the interpretation of such data is complex, and as noted previously, requires expert knowledge.

There are also issues associated with the uneven representation of genetic ancestry in the databases, and therefore the mutations selected for investigation in such studies. Some disease-relevant mutations may be under-represented in the reference population and therefore their significance may be missed for an individual with a different heritage. As others have commented, the important thing ultimately is to focus on how such tests can aid health outcomes for patients, not simply to collect information about genetic anomalies (Lewis and Green 2021).

All of these projects generate a vast amount of information. Genomics is therefore an archetypal "big data" project. This brings into play other ethical questions. Almost inevitably, data sharing involves private companies, in conducting the sequencing itself, and in hosting of the subsequent genomic information. What if those companies suffer a data breach, or choose to sell your data on to a third party? What if a future change of government sees them forcibly requisitioning the genomic health files of their population? Once any information is "in the wild" it is notoriously difficult to recapture.

A fresh spin on the medical use of genomics comes with plans, under discussion at the time of writing, to roll-out routine Whole Genome Sequencing of babies born in the UK. All of the pros and cons we have already discussed remain true, but additional considerations come into play. For example, who gives consent to the procedure? "Clearly the baby is in no position to do so," but ought a parent to be making such a decision on their behalf? And what if the parents themselves disagree? Does the child then get the right to veto their involvement once they reach an age when they are sufficiently competent to make that choice? If they do opt out at that stage, can their data ever be extracted from the relevant databases? What should a clinician report to the family about the results, and when? Is it worse to have a lifetime knowing that a genetic timebomb is ticking? What if advances in treatment mean that something that was not previously considered actionable can now be addressed?

The democratization of genetics

A growing number of people are investigating their DNA not via medical services, but through direct-to-consumer (DTC) companies such as 23andMe, AncestryDNA or one of the burgeoning collections of rival firms. The ethical issues associated with more formal testing (What about incidental findings? Is a finding actionable? Will all people have equal access? And so on) are still present for these customer-initiated tests, but with additional considerations.

As trailed above, a prominent issue associated with DTC tests is the bias in representation of people with different regional heritage within the databases. People with European lineage and Ashkenazi Jews have tended to be over-represented in databases, those from Hispanic and African origin rather less so. This can have significance when it comes to the identification of accuracy of any finding. Although DTC companies are starting to offer whole genome sequencing, the impact of this imbalance in representation is accentuated by the fact that many of the tests still use microarrays. As noted previously, microarrays check for the presence of pre-determined mutations. The particular mutations included in a microarray are often skewed towards information taken from existing databases. There is therefore a significant risk that a mutation which puts a person at risk of a certain disease, may be missed on the grounds that the test is not set up to look for it.

Related to this, there are also potential concerns about the accuracy of the tests. Both sequencing reactions and microchip screens are prone to errors, including false positives and false negatives. To counteract this, well-conducted surveys will reassess the same section of the DNA multiple times, known in the jargon as "depth of coverage". Commercial tests, especially those on the cheaper end of the price range, may not have sufficient depth of coverage to ensure that reports are correct. For example, in one microarray-based study, 4 out of 10 genetically-relevant tests were wrong (Tandy-Connor et al. 2018). There are reports of patients turning up at their doctor asking for elective double-mastectomy to protect them from breast cancer after a DTC test reported that they had a faulty BRCA gene, which gives significant risk of developing the condition when the test was in fact, wrong (Horton et al. 2019).

This leads into another area of ethical concern regarding DTC testing. Very few routinely offer genetic counselling to accompany the return of findings. As discussed, counselling is careful and expensive work, and the cost of providing this service would eat significantly into profit margins. In practice, therefore, a customer with troubling results often goes to their doctor for explanation. In the context of healthcare provided via taxes, such as the NHS, this represents a drain on limited resources. It also presents the physician with a dilemma: should they allow this person to jump the diagnostic queue because they have obtained paid-for information? Might they be sued if they choose not to act on the information they have received? Or equally, might they be sued if they act on information that turns out to be incorrect, but the patient suffers an adverse reaction to the treatment? (Marchant et al. 2021).

Finally, there are issues about consent to secondary uses of genomic data. This is also true for formal genomic analysis, but especially so for the private enterprises. Databases are likely to be made available for at least two purposes: research science and forensics. Genomic repositories offer incredibly rich data for research to understand the genetic basis of disease, but their use is not without controversy. A major issue relates to the consent that people have given when they provided their DNA sample. Did they know that it might be sold on? This information is usually there in the Terms and Conditions of genomic services website, but few people drill down sufficiently to find this out (and their test may even have been bought for them as a present by someone else). Even further, there is the distinct possibility that some of the companies accessing genetic information will be doing so with the intent to commercialise some aspect of testing. If they do, is the subject entitled to a share of the profits?

There is a tension regarding the purpose of genomics—is it for the health of the individual, or for research (albeit research aimed at improving health of wider population)? The latter would perhaps not matter, except that genetic information can reveal things about a person, even if it has apparently been anonymised. For example, a 2018 study estimated that a combination of demographic information and details from commercial genomic sites would allow third-cousins or closer to be identified for 60% of people with European descent (Erlich et al. 2018).

The ability to find relatives matching someone on a DNA database is an attractive prospect for police services around the world. In an early example of this kind of application, Craig Harman was identified as the man who, in 2003, had dropped a brick through the cab of a lorry, causing a fatal crash (Bhattacharya 2004). He was put in the spotlight by a combination of suspect profiling and a partial match to the record of a relative in the UK's National DNA database, collated from samples taken routinely by the police. The most famous case of this kind, however, involved apprehension of the notorious Golden State Killer. During the 1970s and 1980s, a man was responsible for 13 murders and over 50 rapes in California but evaded capture. In 2018, the FBI used the GEDmatch site to match Joseph DeAngelo for those crimes (Guerrini et al. 2018). The following year, FamilyTreeDNA publicised the fact that they had now formally given the FBI access to their 2 million records, justifying the decision by pointing out that police forces were already plumbing genomics archives for leads in cold cases by submitting samples from fictional individuals to look for matches (Haag 2019).

Moving from reading to writing

A different trend in DNA-based science is the capability to intentionally change DNA within cells, a process known as genome editing. When

Jennifer Doudna and Emmanuelle Charpentier published a paper with the underwhelming title *A Programmable Dual-RNA–Guided DNA Endonuclease in Adaptive Bacterial Immunity*, few could have predicted the far-reaching implications (Jinek et al. 2012). There is a hint of the importance in the Abstract, the summary at the start of their article, which concludes by noting that the work "...highlights the potential to exploit the system for RNA-programmable genome editing", but it is unlikely even those authors could have imagined the impact their tool and subsequent revisions has had in the intervening decade.

Doudna and Charpentier were describing a technique that has become known as CRISPR genome editing (pronounced "crisper"), for which they received the Nobel prize in 2020. The trick has been to adapt a system originally used by bacteria to protect themselves by cutting the DNA of invading viruses, into a mechanism for direct alteration of the DNA of an organism at a pre-determined location within its genome.

CRISPR/Cas9, to give its full name, consists of a DNA-cutting enzyme called Cas9, and a guide RNA (a related molecule to DNA). The latter directs the enzyme to the specific place in the genome where it should cut. Unlike previous systems, that required the slow and laborious remaking of their DNA-cutting enzyme every time a new target site was chosen, the advantage of the CRISPR approach is that only the guide RNA needs to be altered. This is a considerably more straightforward and much cheaper proposition. It is a bit like switching to a different computer game by swapping the disc rather than having to replace the console.

On its own, cutting the DNA does not allow for changing of the sequence. Faced with such damage, a cell would naturally try to invoke repair mechanisms, and use the equivalent segment on the second copy of our chromosomes as a template. In order to achieve an intentional change in the sequence, an additional altered copy of that segment is needed, and then it is essential to trick the repair system into using this as the template instead. That may still sound complicated, but it is considerably easier than the previous methods to try and achieve similar goals.

CRISPR has already found widespread use in research, agriculture and therapeutics. Take, for example, potential applications of genome editing in the context of farmed animals. Active projects are investigating the use of CRISPR to achieve a number of different goals (Nuffield 2021). At present, cattle routinely have their horns removed surgically and sheep often have their tails cut off to make husbandry more straightforward. It is now possible to achieve both of these changes genetically. Alternatively, or additionally, it is possible to alter the genes of livestock to introduce other attractive characteristics—the quantity or quality of milk, resistance to a specific disease, or improved heat tolerance, to name but a few.

Elsewhere, CRISPR has been used to engineer mosquitoes in the battle against malaria and other vector-borne diseases such as Zika and Dengue fever. In an ingenious step, the genes for the CRISPR process themselves are incorporated into the mosquito genome alongside any other gene of interest in a "cassette". This leads to a so-called Gene Drive, in which the genes expressed from the cassette cause the equivalent section of DNA on the other chromosome to be cut. To repair the cut, the undamaged copy—the one that includes the gene cassette—is used as the template. This means that it spreads further and faster through the population than would occur naturally. In different strategies, the extra gene in the cassette could be used to induce sterility, or to stop the flies being able to pass on the parasite that actually causes the disease.

It is clever, but it is also controversial. The technology could represent a major step forward against one of the greatest killers on the planet. Over 200 million people per year get infected with malaria, and over half a million die, the vast majority of whom are children (WHO 2021a). There is therefore a strong ethical imperative to do this work. However, to be able to challenge the spread of malaria in affected areas of the world, the engineered mosquitoes are *de facto* going to have to be released into the wild. This raises concerns for environmentalists, who worry that the gene drive system will spread into other unintended species. Because it is crucial that the release goes well, the World Health Organization have published clear guidance on both the efficacy and safety milestones that need to be passed before this can occur (WHO 2021b).

One CRISPR-based modification that did make it into clinical use occurred in January 2022. David Bennett received a transplanted heart that came from a genetically-modified pig. Xenotransplantation, the transfer of organs between different species has been tried before. However, the procedure is inevitably hampered by cell surface markers on the transplanted organ that the recipient identifies as "non-self", triggering immune rejection. The new transplant used a heart from a pig that had been engineered using CRISPR to have ten modifications, making it appear to the immune system as less pig and more human. There were initially encouraging signs that the operation using this "humanised" organ had gone well. However, after two months the patient died (BBC 2022).

Ethical issues relating to genetically modified (GM) xenotransplantation include concerns for the welfare of the source animals (the phrase "donor" is frowned up as it implies an element of choice), worries for the long-term psychological effect of knowing you have a pig's heart and, of course, safety issues for the recipient. The fact that the heart comes from pigs can also trigger religious sensibilities. It may seem that the procedure poses no risk to the wider population. However, the pig genome actually has some viruses embedded within in, so these might also need to be modified to guard against inter-species transfer of infection.

GM Humans

Could CRISPR ever be used to directly modify a human? The answer is not only yes, but it has already been done. In an infamous case, Chinese doctor He Jiankui announced in November 2018, that he had used CRISPR to genetically modify twin girls. He had employed this tool to make a non-functional version of protein called CCR5, which is exploited by HIV to enter cells. In theory, therefore, the girls ought to be protect against infection with the virus.

The news scandalised the scientific community. There remain significant concerns about "off target" effects, in which the CRISPR-associated enzyme cuts the DNA at other unintentional places in the genome, with unknown consequence. This alone made his work premature. Furthermore, the CCR5 protein isn't just sitting around waiting to be a conduit for the virus to enter cells—it has a job of its own modulating cell signals during normal immune responses to infection. He has effectively removed this element of protection for the girls (see Greely 2019 for a thorough critique of what was, in effect, an *in vivo* experiment).

Following the global outcry, the Chinese authorities sentenced him to three years in prison (he was released in April 2022). This remains the only known example to date of someone intentionally modifying the hereditable human genome, apart from a third baby, also modified by his team, for which no details have been made public (indeed the work for which he was sent to prison has never been formally published). However, since CRISPR use does not require specialist training and equipment, it is likely that others may already be attempting to perform similar alterations somewhere in the world, even if current regulations prohibit it in many countries.

Beyond the technical and safety issues, trying to exploit genome editing to remove some features considered undesirable raises the spectre of eugenics, and previous attempts to eliminate "deleterious" characteristics from the human gene pool—in those cases by deciding who had the right to reproduce, and with whom. This also leads to a discussion on the extent to which parents have a moral right to intervene in gene-based aspects of their children, particularly those that will affect not only them as individuals, but also all their descendants. If this trend was to become fashionable and the technology routine, there is a risk that humans lose their diversity, conforming to a socially-determined norm. Shutting the door to all genetic modifications to avoid ethical quandaries could mean missing out on important health improvements, but this reinforces the need for interventions only to be made when the outcomes are better understood.

Before returning to the potential to genetically modify humans, it is important to note that interventions right at the start of life, as exemplified by Jiankui's work, remind us that aspects of reproduction are another "hot topic" in bioethics—both in the sense of being currently active, but also being controversial.

Ethical implications of development in human reproduction

New beginnings

The implications of moves to roll out routine genome sequencing of newborns have already been discussed, and latterly also considered the potential to introduce intentional changes to the DNA of embryos. These are just two of a number of current innovations related to the beginning of life.

In vitro fertilisation (IVF) is not itself a new technology (Louise Brown, the first IVF baby, was born in 1978). It is fair to say, however, that the majority of new areas of development, and controversy would not be feasible without IVF. Several of these involve gamete donation and/or storage.

Sperm donation *per se* does not necessarily involve IVF, there were methods of artificial insemination feasible before these more sophisticated approaches. In passing, however, it is just worth noting two ethical points about sperm donation. Firstly, there is the issue of anonymity. Several jurisdictions maintain a policy of confidentiality about the identity of the donor, whilst others have mechanisms for revealing this information. The rise of DTC genetic testing now undermines the plausibility of anonymity (Harper et al. 2016). Secondly, as more information emerges about the genetic heritage of children conceived by IVF, it has become clear that some men have fathered considerable numbers of offspring. This raises the possibility that their genomes are significantly over-represented in the population at large and, without any safeguards, it is feasible that half-siblings might become a couple in their own right and thereby be participants in inadvertent consanguineous conceptions. Regrettably, it has also become clear with alarming frequency that the biological father of a child conceived at a fertility clinic turns out to be the man running the clinic, having substituted his own sperm for that of a different donor or indeed the partner of the woman seeking assistance.

If sperm donation has been possible for a while, egg donation is a rather newer phenomenon. Not only are far fewer eggs (oocytes) produced, but their harvesting is more intrusive and places a not insignificant physiological burden on the donor. Coupled with this, the capability to safely freeze and rethaw eggs required the evolution of cryopreservation protocols; the first live birth following the use of vitrification in egg freezing was reported in 1999 (Kuleshova 1999).

Aside from sharing some of the issues of anonymity in an age of genomic data, the other ethical dimension to recent developments in regard to egg freezing have primarily related to the rise of cryopreservation for non-medical reasons, sometimes called "social" egg freezing. Unlike traditional services provided for those struggling to conceive, social egg freezing is targeted at presumably healthy women who are looking to safeguard their fertility against the inevitability of age-related decline (Baldwin 2019). Terms such as "insurance", "investment" and "banking" are commonly employed, emphasising that this is looking towards the woman's opportunity to have genetically-related children in the future. As is so often the case, there is uneven opportunity—the costs of both egg freezing and subsequent IVF procedures preclude many women from participating, reinforcing the division between the haves and the have nots.

There has been a growth in companies, particularly entrepreneurial tech giants such as Apple and Facebook offering egg freezing as a perk. This is not without its own difficulties. For example, what happens if and when a woman's employment at the company ends (either through their choice or hers)? Is there an ongoing commitment to the storage? In many ways this is a wrong answer to the genuine problem, namely that workplaces are inadequately geared up to support women, especially those embarking on having a family (Baldwin 2019).

DIY gametes

The landscape regarding gamete sources may be set to change in much more dramatic ways. These exploit the growing understanding of how to cause normal body cells, notably skin cells (for ease of accessibility) back into more malleable "induced pluripotent stem cells" (iPS cells). There have been significant recent breakthroughs in which human iPS cells have been re-programmed to become gametes and then developed on to become early embryo-like "blastoids" (at which point the work was stopped for ethical reasons) (Liu et al. 2021, Yu et al. 2021). If *in vitro* gametogenesis becomes an established procedure, then eggs or sperm could be made from anyone's cells. As well as other applications, the possibilities then stretch to a same-sex couple having a child that is genetically derived from *both* of them. This is a step on from current scenarios in which lesbian couples sometimes share parenting by one providing the egg and their partner carrying the subsequent pregnancy. Gay men would still require a surrogate, although other development in ectogenesis and the feasibility of gestation occurring in an artificial womb might even put pay to that (Aguilera-Castrejon et al. 2021).

The challenges of post-humanism

Re-defining the future of humanity

How close do we stand to the possibility of fundamental alterations to the human species? As well as offering the potential to engineer genes so that neither the newborn nor their descendants suffer from a given genetic disease, might the same genome editing methods be used to introduce enhanced features, over and above those of existing humans?

Although the notion of "superhumans" has been a staple from ancient mythology through to contemporary science fiction, we stand at a point where the technologies necessary to achieve this are on the cusp of plausibility. This is not just true in regard to genome editing, but also other "biotransformative technologies" (Porter 2017) such as pharmacological and electromechanical interventions. The latter may be less developed than either the genomic or pharmaceutical approaches but significant recent breakthroughs have occurred in this field. Additionally, as noted at the outset, ethics has far too frequently been left playing catch-up so it certainly not premature to reflect on those aspects.

Before doing so, it is important to reflect on pharmaceutical interventions. There is growing interest in medications that might contribute to enhanced brain performance. Such cognitive enhancers, sometime referred to as smart drugs or nootropics, are initially compounds that might be prescribed to alleviate particular symptoms but are considered to offer benefits for "off label" use. These include Modafinil (Provigil) which promotes wakefulness when fighting narcolepsy or shift-work sleep disorder, but has proven popular with students to improve attention and alertness, especially in the context of "pulling an all-nighter" to write an essay shortly before the deadline. Similarly, Donepezil (Aricept) officially used in treatment of mild to moderate Alzheimer's disease is sometimes taken to aid retention of training information. Other enthusiasts seek to achieve similar ends via the use of electrical brain stimulations such as transcranial magnetic stimulation and transcranial direct current stimulation (Yazdi 2020).

This crosses over into the "shallow end" of more electromechanical interventions, in the sense that these attempts at brain modification involve external, i.e., non-invasive, stimulation. Placing a device entirely or partially within the body represents a significant step further. Some of these, putting magnets or radio frequency identification (RFID) chips under the skin are relatively low-risk and procedures might be conducted by body modification enthusiasts, sometimes known as "grinder biohackers" for relatively whimsical reasons. RFIDs may have a more serious purpose. These could include documenting medical records, regulating delivery of a medicine, or activating mobility-enhancement devices for the physically disabled. They come, however, with concerns about privacy and data security, and the potentially harmful outcomes if they were cloned, counterfeited or manipulated remotely in other ways (Juels 2006).

A further threshold, both in terms of potential significance and health risk, is crossed when we move on to contemplate direct interaction with the nervous system of a recipient. The capability to achieve functional and stable brain-machine interfaces (BMIs) is being advanced by such innovations as the miniaturisation of components and development of more biocompatible materials. Hitherto, most of the work done in this area has been directed towards therapeutic purposes, such as the returning of limb control to individuals who have suffered traumatic spinal injury (e.g., Collinger et al. 2013, Salas et al. 2018, Flesher et al. 2021).

As with so many of these innovations, however, the same approaches could be adapted to enhancement goals. This might be, for example, connection of the body to a super-strong exoskeleton to equip future infantry. Other applications might allow a user to have "augmented cognition", able to access information from an external server direct to the brain. Such technology-enhanced humans would fit into the category of cyborgs since they are at source still organic in nature.

Approaching the same interface from the other end would be intelligence-augmented robots, or androids, which are entirely machine. Whether or not such a being is worthy of personhood and autonomy has long been the substance of science-fiction including, but in no way exclusively, *Blade Runner* (1982), *Bicentennial Man* (1999), *ExMachina* (2014) and the British television series *Humans* (2015–2018). Consideration of these issues has also received more formal academic attention (e.g., Hubbard 2011).

Although the definitions are contested, what we are considering here are notions of transhumanism, in which the natural human form remains, but with enhanced characteristics, and posthumanism, in which such fundamental changes have taken place that the core essence of what it means to be human has been significantly altered, so that we are effectively talking about a different species (Bostrom 2005). Various terms are used for the range of enhancements people may consider. Authors speak of those that are "normal-range enhancements", such as moving somebody from the lowest quartile of human intelligence to the highest. And "normality-transcending" enhancements, for example the capacity to live much longer than usual (e.g., Kahane et al. 2016). Others distinguish "extraordinary augmentations", that is the ability to perform some normal human function such as running peculiarly fast and "radical augmentations", things that natural humans could not do, such as to fly (e.g., Shook and Giordano 2016).

Whatever the scope of enhancements, and however they are branded, it will come as no surprise to learn that attempts to make such alterations to humanity face opposition. Those raising concerns about the impact of the rise of transhuman and posthuman forms are frequently termed "bioconservatives".

A variety of arguments are used to support the bioconservative view that we should not be travelling in this direction. These tend to be based on deontological arguments such as concerns about the challenge to dignity, the unfairness of some families being able to afford enhancement programmes whilst others cannot, and the potential threat to "ordinary" humans posed by the emergence of some form of enhanced human 2.0. Philosopher Michael Sandel (2004) has a slightly different take; the pursuit of perfection both undermines our contentment with the existing human reality and fractures the relationship between a parent and their enhanced child.

Anticipating counter-punches, Sandel (2004) notes that a transhumanist pointing out that intervening genetically to improve your child is no different to providing them with extensive and expensive training programmes to achieve greater academic or sporting prowess is not so much a defence of technological enhancement as it is a criticism of contemporary "hyperparenting". The more we reject the inherent randomness of the "given", not only do we demonstrate our hubris, we open ourselves up to accusation of blame by others (for example, the ungrateful offspring) and to the angst of knowing that we literally are responsible for the success of failure of our chosen intervention. Sandel (2004) further argues that our capacity to appreciate and admire the feats of others, say for sporting achievement, is undermined by the knowledge that they have been selected to excel in that specific arena.

Looking in the opposite direction, that is to say a transhumanist critiquing bioconservatism, Julian Savulescu and colleagues note the centrality of the Human Nature Objection (HNO) to arguments put forward by the latter (Kahane et al. 2016). At its core the HNO states "they are not us" (this is, incidentally, accused of being part of a broader "speciesism" which we will not unpack here). The identification with "us" can be driven by intrinsic value a person or thing possesses ("particular valuing") or via the special connection of that item to the evaluator ("personal valuing"). Kahane et al. (2016) take the example of a musician who has a peculiar fondness for a damaged violin that ought logically to be replaced, because it was the very instrument on which they first learned to play. They go on to state that the HNO is at its strongest when it aligns with personal valuing ("it's *our* nature") rather than appealing to intrinsic value. Remember, though that they are not bioconservatists, and they ultimately determine that this is insufficient to countermand the merits of some forms of human enhancement.

Conclusions

In this chapter, some of the most significant current and emerging trends in biomedicine were surveyed and ethical issues associated with these developments were brought to light. The initial focus was on aspects of both genome sequencing and genome editing, since these seem to be travelling at greatest pace and to offer the most plausible changes to humanity. Next, other issues associated with the beginning of life were considered, and how a range of different yet closely associated innovation are broadening the horizons of potential ways to conceive new humans. Finally, both the biological and electromechanical advances that may give rise to beings that are either superhuman, or beyond definition as *Homo sapiens*, were discussed.

The premise throughout has been that early dialogue with a well-informed public in essential in shaping the ethical boundaries—just because something can be done, doesn't necessarily mean that it should. Fear of the consequences of a change ought not always to hold us back, but it ought always to trigger careful reflection. Along the way, the tension between therapy and enhancement was highlighted. Frequently an innovation targeted at alleviating the suffering of an individual generates methodologies that can readily be turned towards enhancement. This can be beneficial but, once again, should make us pause and consider the implications at the earliest juncture. It can be argued, in particular, the importance of encouraging the population to be as genetically-literate as possible. The hope is that this chapter will contribute to that process.

References cited

Aguilera-Castrejon, A., B. Oldak, T. Shani, N. Ghanem, C. Itzkovich, S. Slomovich et al. 2021. Ex utero mouse embryogenesis from pre-gastrulation to late organogenesis. Nature 593: 119–124.

Ali, H.R., O.M. Rueda, S.F. Chin, C. Curtis, M.J. Dunning, S.A.J.R. Aparicio and C. Caldas. 2014. Genome-driven integrated classification of breast cancer validated in over 7,500 samples. Genome Biol. 15: 431.

Baldwin, K. 2019. Egg Freezing, Fertility and Reproductive Choice: Negotiating Responsibility, Hope and Modern Motherhood. Emerald Publishing, Bingley.

BBC. 2022. Man given genetically modified pig heart dies. https://www.bbc.co.uk/news/health-60681493.

Bhattacharya, S. 2004. Killer convicted thanks to relative's DNA. News Scientist (20th April 2004). https://www.newscientist.com/article/dn4908-killer-convicted-thanks-to-relatives-dna.

Bostrom, N. 2005. In defense of posthuman dignity. Bioethics 19: 202–214.

Collinger, J.L., B. Wodlinger, J.E. Downey, W. Wang, E.C. Tyler-Kabara, D.J. Weber et al. 2013. High-performance neuroprosthetic control by an individual with tetraplegia. The Lancet 381: 557–564.

Erlich, Y., T. Shor, I. Pe'er and S. Carmi. 2018. Identity inference of genomic data using long-range familial searches. Science 362: 690–694.

Flesher, S.N., J.E. Downey, J.M. Weiss, C.L. Hughes, A.J. Herrera, E.C. Tyler-Kabara et al. 2021. A brain-computer interface that evokes tactile sensations improves robotic arm control. Science 372: 831–836.

Greely, H.T. 2019. CRISPR'd babies: human germline genome editing in the 'He Jiankui affair'. Journal of Law and the Biosciences 6: 111–183.

Grenz, S.J. 1998. The Moral Quest. InterVarsity Press, Leicester.

Guerrini, C.J., J.O. Robinson, D. Petersen and A.L. McGuire. 2018. Should police have access to genetic genealogy databases? Capturing the Golden State Killer and other criminals using a controversial new forensic technique. PLoS Biol. 16: e2006906. https://doi.org/10.1371/journal.pbio.2006906.

Guttmacher, A.E. and F.S. Collins. 2003. Welcome to the genomic era. N Engl. J. Med. 349: 996–998.

Haag, M. 2019. FamilyTreeDNA Admits to Sharing Genetic Data With F.B.I. New York Times (4th February 2019). https://www.nytimes.com/2019/02/04/business/family-tree-dna-fbi.html.

Harper, J.C., D. Kennett and D. Reisel. 2016. The end of donor anonymity: how genetic testing is likely to drive anonymous gamete donation out of business. Human Reproduction 6: 1135–1140.

Horton, R., G. Crawford, L. Freeman, A. Fenwick, C.F. Wright and A. Lucassen. 2019. Direct-to-consumer genetic testing. BMJ 367: l5688.

Hubbard, F.P. 2011. Do androids dream: personhood and intelligent artifacts. Temple Law Review 83: 405–474.

Jinek, M., K. Chylinski, I. Fonfara, M. Hauer, J.A. Doudna and E. Charpentier. 2012. A programmable Dual-RNA–guided DNA endonuclease in adaptive bacterial immunity. Science 337: 816–821.

Juels, A. 2006. RFID security and privacy: a research survey. IEEE Journal on Selected Areas in Communications 24: 381–394.

Kahane, G., J. Pugh and J. Savulescu. 2016. Bioconservatism, partiality, and the human-nature objection to enhancement. The Monist 99: 406–422.

Kuleshova, L., L. Gianaroli, C. Magli, A. Ferraretti and A. Trounson. 1999. Birth following vitrification of a small number of human oocytes: Case Report. Human Reproduction 14: 3077–3079.

Lewis, A.C.F. and R.C. Green. 2021. Polygenic risk scores in the clinic: new perspectives needed on familiar ethical issues. Genome Med. 13: 14.

Liu, X., J.P. Tan, J. Schröder, A. Aberkane, J.F. Ouyang, M. Mohenska et al. 2021. Modelling human blastocysts by reprogramming fibroblasts into iBlastoids. Nature 291: 627–632.

Marchant, G.E., M. Barnes, E.W. Clayton and S.M. Wolf. 2021. Liability implications of direct-to-consumer genetic testing. *In*: Cohen, I.G., N.A. Farahany, H.T. Greely and C. Shachar (eds.). Consumer Genetic Technologies: Ethical and Legal Considerations. Cambridge University Press, Cambridge, England.

Nuffield. 2021. Genome editing and farmed animal breeding: social and ethical issues. Nuffield Council on Bioethics. https://www.nuffieldbioethics.org/publications/genome-editing-and-farmed-animals.

Nuffield. 2022. New infographic highlights ethical issues 'on the horizon' in biology and medicine. Nuffield Council on Bioethics. https://www.nuffieldbioethics.org/news/new-infographic-highlights-ethical-issues-on-the-horizon-in-biology-and-medicine.

Porter, A. 2017. Bioethics and transhumanism. Journal of Medicine and Philosophy 42: 237–260.

Salas, M.A., L. Bashford, S. Kellis, M.J.H. Jo, D. Kramer, K. Shanfield et al. 2018. Proprioceptive and cutaneous sensations in humans elicited by intracortical microstimulation. eLife 7: e36137.

Sandel, M. 2004. The case against perfection. The Atlantic. https://www.theatlantic.com/magazine/archive/2004/04/the-case-against-perfection/302927.

Shook, J.R. and J. Giordano. 2016. Neuroethics beyond normal: Performance enablement and self-transformative technologies. Cambridge Quarterly of Healthcare Ethics 25: 121–140.

Tandy-Connor, S., J. Guiltinan, K. Krempely, H. LaDuca, P. Reineke, S. Gutierrez et al. 2018. False-positive results released by direct-to-consumer genetic tests highlight the importance of clinical confirmation testing for appropriate patient care. Genetics in Medicine 20: 1515–1521.

Watson, J.D. and F.H.C. Crick. 1953. Molecular structure of nucleic acids: a structure for deoxyribose nucleic acid. Nature 171: 737–738.

WHO. 2021a. Malaria. World Health Organization. https://www.who.int/news-room/fact-sheets/detail/malaria.

WHO. 2021b. Guidance framework for testing genetically modified mosquitoes. Second edition. World Health Organization. http://apps.who.int/iris/bitstream/handle/10665/341370/9789240025233-eng.pdf.

Yazdi, P. 2020. What is tDCS & Are DIY Devices Safe? SelfHacked. https://selfhacked.com/blog/tdcs-diy-devices-safety.

Yu, L., Y. Wei, J. Duan, D.A. Schmitz, M. Sakurai, L. Wang et al. 2021. Blastocyst-like structures generated from human pluripotent stem cells. Nature 591: 620–626.

Chapter 9

Roles and Challenges for Clinical Ethics Committees and Clinical Ethics Consultation Systems

*María Bernardita Portales Velasco** and *Juan Pablo Beca Infante*

Introduction

The major technological and scientific developments of the last 80 years have resulted in enormous advances in medicine, including an ever-increasing body of knowledge about the diagnosis and treatment of multiple diseases and health conditions. This evolution, which continues apace, prompted the need to devise a medicine and a medical science grounded in values. As early as the 1970s, this necessity led to the development of a new discipline or interdiscipline: bioethics. The emergence of this discipline, which reflects on issues such as the boundaries of medicine, its objectives, how to respect the patient's autonomy, or who will safeguard the good of the patient, was also influenced by the social and cultural changes that characterized those times. More awareness was attained of the moral plurality that typifies democratic societies, where each person has their own scale of values, with their own beliefs and preferences in their culture and context. In the health care field, the patient has been recognized as a moral agent who has an opinion

Universidad del Desarrollo, Chile.
* Corresponding author: bportales@udd.cl

and who can make decisions on his own health. This position has been cemented through several declarations, letters, and regulations, the first of which was the Patient's Bill of Rights, issued by the American Hospital Association in 1972. This recognition of the patient as a rights bearer who must be considered and respected when making decisions replaced the paternalistic model that had characterized medicine since the times of Hippocrates, leading to more horizontal physician-patient relationship models. Along with this change, the clinical relationship was extended, no longer being limited to the physician and the patient; nowadays, it is accepted to encompass an interdisciplinary team of professionals and the patient's family, within a context that includes administrative and health insurance factors, among others.

Despite the recent medical progress, decision-making continues to be fraught with uncertainty and complexity: prognosis is uncertain and case-dependent, available resources differ across areas and hospitals, treating teams can have a wide range of experience levels, and it is necessary to take into account the patient's preferences and wishes. Decisions must be made without forgetting the obligation to safeguard the patient's good, informed by medical judgment but considering all the factors at play, which include the equitable distribution of the available health care resources (González-Bermejo et al. 2020). Furthermore, professional over-specialization sometimes results in fragmented patient care, generating distrust and discontent. In such contexts, value-related conflicts emerge which do not have a single right solution. Therefore, it is necessary to embark on a deliberative, rational, and ethically grounded process whereby, respecting the patient's rights, we seek courses of action conducive to attaining the greatest good, that is, the best decision for each individual patient. All these reasons encouraged health care institutions to form clinical ethics committees (CECs) and, whenever possible, clinical ethics consultation services, as both systems help to devise mechanisms that protect patients' rights, fostering shared decision-making and improving the quality of the care delivered (Portales 2022).

In this chapter, we offer a brief overview of the origins and functions of clinical ethics committees. Data from the United States, Spain, and Chile will be referenced. We will examine the emergence of clinical ethics consultation services and explore how, in our experience, such services operate to provide timely solutions to the ethical conflicts that emerge when delivering health care. The chapter ends with a reflection based on the experience of its authors regarding the challenges and new roles of CECs and clinical ethics consultation systems.

Definition and origins of clinical ethics committees

Clinical ethics committees is a multidisciplinary and deliberate body that furnish clinical settings with a formal ethical perspective (Post and Blustein 2021). These committees, of a consultative nature, serve users and professionals at health care institutions, contribute to solving the ethical problems that emerge when delivering care, and issue recommendations based on an ethical and clinical analysis of each case or situation, thus helping to improve service quality (Portales 2022).

Clinical ethics committees emerged as early as the 1960s, in the USA. In 1960, Dr. Belding Screibner created the first hemodialysis center for renal outpatients (Seattle Artificial Kidney Center). However, since hemodialysis was a scarce resource at the time, it was necessary to identify the patients who would benefit the most from this technique. For this reason, a multidisciplinary committee was created. Its members—a physician and members of the community—used moral in addition to medical criteria to select hemodialysis candidates. This body, also known as "The Life and Death Committee" or "God's Committee", was short-lived because dialysis became more easily accessible (Bravo 2012). Although it played a relevant part in the history of CECs and is sometimes regarded as one of their precursors, this group did not promote the creation of other committees and "failed to develop an adequate set of criteria" (Ferrer 2003:27).

In the 1970s, after criteria for determining brain death were published by Harvard University in 1968, and as a result of ethical conflicts encountered in neonatology units related to the protection of children with disabilities, care ethics committees were created in several US hospitals (Bravo 2012). In the late 1970s, committees were also set up to analyze the cases and ethical conflicts generated by *in vitro* fertilization and embryo transfers, two procedures that were being developed in England (Bravo 2012).

In 1975, pediatrician Karen Teel published an article in the *Baylor Law Review* where she stated that ethics committees could assist physicians faced with ethical conflicts when treating pediatric patients; however, her proposal caused some confusion due to the assertion that these committees should be tasked with establishing the patient's prognosis (Myers and Lantos 2014). In 1976, a report issued by the Optimum Care Committee at Massachusetts General Hospital describes a mode of operation similar to that of current CECs and outlines how it offered advice to physicians (Pose 2016).

In the 1980s, a report issued by the President's Commission for the Study of Ethical Problems in Medicine and Biomedical and Behavioral Research (1983) recommended the creation of clinical ethics committees "in order to share decisions in clinical settings and thus ease the responsibility of physicians in cases posing ethical problems" (Pose 2020:8). In 1984, the "Baby Doe Rules" were promulgated, and the United States Department of Health and Human Services (DHHS) and the American Academy of Pediatrics recommended the creation of Infant care review committees in hospitals.

Given their relevance for the history of CECs, two cases are worth examining in more detail. The first of them, which undoubtedly encouraged the establishment of CECs, is that of Karen Ann Quinlan. Indeed, the decision of the New Jersey Supreme Court in 1976 "represented the first time that, in response to a suit, a court recommended the intervention of a CEC, as it concluded that the resolution of moral problems associated with clinical practice must be solved by a CEC, without needing to initiate legal procedures" (Bravo 2012:245).

On April 15, 1975, Karen Ann Quinlan, a young American woman, went into cardiac arrest at a party after consuming alcohol and barbiturates. In a coma and weighing 50 kilos, she was taken to Newton Memorial Hospital in New Jersey, where she was connected to a ventilator (Velasco-Sanz 2005). Karen Ann's parents were a Catholic couple who had adopted the girl when she was only a few weeks old. After Karen Ann had been connected to a ventilator for 3 months, and seeing her gradually deteriorate, her parents requested that she be disconnected from her ventilator, as her physicians had informed them that there was no reasonable hope of recovery. "The doctors refused, arguing that executing the family's order constituted 'a moral issue', since such an act would be tantamount to homicide" (Velasco-Sanz 2005:161). The family filed this request before the courts in November 1975; on March 31st, 1976, the New Jersey Supreme Court ruled that Karen Ann Quinlan's ventilator could be lawfully removed. Finally, on May 14th, 1976, Karen Ann was disconnected from her ventilator and taken to her home. There, she lived for 9 years in a vegetative state and was given nasogastric feeding until she died of pneumonia in 1985.

The other case that had a major impact on the origin of CECs was that of Nancy Cruzan, who was involved in a traffic accident in Missouri in 1983. This married young woman, aged 24, was reanimated and taken to a hospital, where she was diagnosed as being in a persistent vegetative state. After one month of hospitalization, her father authorized a gastrostomy (Colby 2019). Later on, the young woman became able to breathe spontaneously. After 4 years, her parents requested that her feeding tube be removed, which was rejected by her physicians. This case was heard by the Supreme Court of the United States in 1990. The Court ruled in favor of accepting the request filed by the woman's parents, which resulted in the Patient Self-Determination Act. This act establishes the rights of competent patients to reject a treatment or medical procedure, allowing duly authorized third parties to make decisions on behalf of incompetent patients (Martínez 2007).

In 1992, the Joint Commission on Accreditation of Healthcare Organizations established that institutions seeking accreditation would need to have mechanisms in place to solve moral issues. For its part, UNESCO, in the Universal Declaration on Bioethics and Human Rights (2005), "urges States to adopt all the necessary measures, be them legislative, administrative, or of any other type, to protect people's rights and dignity, especially through the creation of independent, multidisciplinary, and pluralistic ethics committees" (González-Bermejo et al. 2020:207), thus encouraging the establishment of ethics panels and the delivery of the relevant training in bioethics for health care professionals (Gómez 2015). The requirement set by the Joint Commission on Accreditation of Health Care Organizations continues to play a major role in the establishment of and access to clinical ethics committees or Ethics Services at the health care institutions that have been granted this accreditation.

Thus, the number of clinical ethics committees has gradually increased in the USA, while their development has been more variable in other countries. These CECs, also known as Bioethics Committees, have become more common in hospitals—hence the name Hospital Ethics Committee, but they are not limited to these settings, since they have also been established in Primary Health Care Centers.

In Spain, the first clinical care committee was created in 1976 at San Juan de Dios Hospital in Barcelona. This body was mostly composed of physicians and nursing professionals who periodically gathered to ponder ethical issues (Abel 2006). "But it was only in the 1990s that their creation was most strongly encouraged, as a result of the publication of the Order of December 14th, 1993, on the accreditation of CECs in Cataluña, Circular 3/1995 on the creation and accreditation of CECs in the INSALUD system, and Decree 143/1995 of the Basque Government" (González-Bermejo et al. 2020).

In Chile, clinical ethics committees can be traced back to the CEC at Luis Calvo Mackenna Hospital, which was formally established in 1990. Technical General Norm No. 2 of 1994 specified the roles, powers, and modes of operation of the Medical Ethics Committees of Health Care Services, which oversaw both clinical and research activity. In 2005, this norm was modified, separating the functions of clinical ethics committees from those of their research ethics counterparts. Lastly, Law No. 20.584, which "Regulates the rights and duties of people in connection to actions concerning their health care", promulgated in April 2012, explicitly references these CECs (articles 17 and 18). On that same year, a specific set of regulations was developed for the establishment and operation of these committees in Chile (Decree 62, 2012).

Functions and characteristics of CECs

Clinical ethics committees are created in order to advise those who must make decisions in health care settings. Here, through a deliberative process, these groups of people address pressing clinical care needs in their organizations. As previously noted, decision-making in the health care field involves health care professionals, patients, and/or their representatives, all of whom can request CEC support; however, personnel in charge of institutional policies, management, and administration of a health care center can also benefit from CEC assistance (Fernández Letamendi 2016). For this reason, clinical ethics committees have three main functions: consultative, normative, and educational:

- The *consultative function* refers to the analysis of the cases of patients of any age whose treatment or associated decision-making processes lead to ethical doubts or conflicts. Traditionally, this has been the most common function of these committees. This consultative function facilitates the identification and resolution of value-related conflicts, with the CEC issuing ethically-grounded recommendations. Requests for advice can be filed by any health care professional, including physicians and non-medical professionals, the patient, and/or his/her family. The formal characteristics of these requests and the procedure employed will depend on the mode of operation of each CEC.
- The *normative function* consists in the creation and proposal of protocols or guidelines derived from the analysis of similar cases or frequent situations observed in the health care center that hosts the CEC. This normative function also encompasses the committee's potential contributions to institutional protocols and policies.
- The *educational function* is conducted through bioethics training programs for committee members as well as bioethics teaching activities aimed at all the institution's personnel, also including community outreach initiatives. Bioethics training programs enable CEC members to offer better support for the committee's recommendations or norms, strengthening their deliberations and ethical-clinical analyses.

The following are functions that fall outside the remit of a clinical ethics committee. CECs cannot judge the actions of the professionals working at their host institution, since this is part of the Auditing Departments' duties. They cannot issue judgments on the professional ethics of the personnel, since this task is among the responsibilities of an ethics tribunal. CECs are not a resource or a body intended for the legal defense of health care personnel. Lastly, it is worth noting that a person requesting assistance from a CEC is still personally responsible for his/her actions. Recommendations issued by CECs as part of their advisory and consultative functions are non-binding, which means that they are not obligations, only suggestions (Gómez 2015).

Clinical ethics committees must systematize their functioning and specify their composition (membership). The number and characteristics of CEC members can vary depending on the relevant regulations, but it is important that they be interdisciplinary and intergenerational. This composition, which is one of the main contributions of these committees, makes it possible to encompass a variety of well-supported views, notions, and opinions representing multiple moral perspectives found in the context of each CEC (Altisent et al. 2019, Morlans 2021).

Even though clinical ethics problems are common in health care practice, and acknowledging the importance and contributions of CECs, the literature shows that these committees examine a limited number of cases. Ellen Fox, based on a survey published in 2021, reports that an average of 3 cases per year are submitted to clinical ethics advisory bodies in US hospitals. Furthermore, 14% of the hospitals had no cases submitted to CECs during the year prior to the survey and only 7.1% of these bodies analyzed more than 60 cases (Fox et al. 2021). In Chile, an average of 6 to 7 cases are presented to CECs annually (Oficina Bioética MINSAL 2019).

The low number of cases examined—often the most extreme ones—may be a result of the operation of CECs, which have a fixed meeting schedule and cannot address clinicians' needs in a timely manner; furthermore, it is often difficult to access CECs or there is ignorance of how to make the request. However, CECs may also be underutilized because health professionals fail to identify that the problems they face involve ethical issues and are unaware of the benefits of receiving ethically-grounded recommendations. Another possible reason is that professionals, patients, and/or relatives may feel intimidated when they present their situation at a plenary committee meeting; mistrust the confidentiality of their interactions with CEC members, or fear being judged by the committee. Lastly, they may feel discouraged due to the time needed to prepare the presentation of the case and may not know exactly what information the committee needs to analyze their clinical ethics case (Portales 2022).

To make an actual contribution to professionals and users, clinical ethics committees require institutional support and acknowledgment. In this regard, Rogelio Altisent notes that CECs "must inspire trust, welcoming those who turn to them looking for advice and preventing them from feeling judged" (Altisent et al. 2019:24). Altisent lists several requirements for the proper operation of CECs, the first of which is that they must be accessible to those who need it. Accessibility requires that the CECs must remain visible to the professional and user community and also requires the implementation of more current forms of access that facilitate consultation, such as the provision of telematic channels. Committees must also be agile, combining the necessary calm needed for deliberation with adequate time to give a response, without long delays, so that they can be useful to those requesting assistance. In addition, they need to be efficient, because it must be possible to justify that the resources invested in the CEC are necessary to carry out its functions. Finally, they have to be constantly self-evaluated, incorporating quantitative aspects regarding the operation, meetings, attendees, cases presented, etc. and qualitative factors. This will allow identifying strengths, weaknesses to work on, and making comparisons with other CECs (Altisent et al. 2019).

Clinical ethics consultation services or systems

In order to resolve clinical cases involving ethical issues promptly, several institutions have developed clinical ethics consultation services or systems offered by a suitably trained CEC member or subgroup. Therefore, ethical consultation services can be delivered by (1) the ethics committee in its meetings, (2) an individual consultant, or (3) a small group of consultants. Juan Pablo Beca defines clinical ethics consultation—aimed at improving the quality of health care services—as "a contribution to the analysis and identification of clinical ethics problems in individual cases to ease the resolution of value-related conflicts, with the participation of the people involved in the case, leading to the issuance of well-grounded recommendations" (2012:254).

The ethical consultation service model implemented by Clinical Ethics Committees predominates in Spain and the rest of Europe (Fernández 2016). A survey conducted by Ellen Fox (published in 2021) shows that, in most US hospitals, consultation services are chiefly provided by a small group of CEC members (65.1%), with a smaller share being delivered by the whole committee in plenary meetings (16.3%) or by individual consultants (18.6%), (Fox et al. 2021). Carlos Pose (2017) argues that "the role of individual clinical ethics consultants will grow over time in clinical settings, but without substituting clinical ethics committees; rather, they will energize them or operate as mediators between the clinical ethics problems of patients or professionals and clinical care committees" (Pose 2017:72).

Consultation services and the concrete functioning of committees can differ among health care centers according to local cultures, resources, or specific experiences, among other factors. For instance, some centers may hold online consultation sessions, allowing the people involved who cannot attend in person to have a say in the decision-making process. This situation was especially common during the COVID-19 pandemic, which also prompted the establishment of CEC subgroups to aid in complex decision-making processes involving patients with this diagnosis. Despite potential differences, these services have made it possible to bring bioethics closer to the patient's bed. In this regard, George Agich points out that "clinical ethics consultants are right at the scene; they are involved in the cases which they are asked to assess. They communicate with patients, relatives, clinicians, nurses, and other health care professionals about the case" (2005:10), thus bringing bioethics closer to the real situation being experienced and the context where support is being requested (Portales 2022).

These consultation systems have made it possible to contextualize each case more thoroughly, offer direct support to patients and their family, and aid the treating physicians in the ethical analysis of problematic decisions, thus reducing their moral stress, increasing their ethical sensitivity, and improving their comprehension of clinical ethics issues as well as their bioethics knowledge (Portales and Beca 2021).

As previously noted, it should be possible for clinicians and other health care professionals, the patient, his/her family, or his/her legal representative to request clinical ethics consultation services. Formally, consultations are assumed to be referrals and should ideally be free of charge for patients, since they support both the professional team and the patient when it is necessary to make complex decisions. As with all referrals, there is one key question that often reflects the ethical problem that must be resolved and that prompts this request. Even though other ethical problems can be identified during the deliberative process, it is relevant for the consultation report to focus on answering the question that prompted the referral; therefore, the search for an answer to this core question should guide the committee's deliberation on courses of action and recommendations.

How do clinical ethics consultants work?

This section covers a mode of functioning adopted by the authors of this chapter, which is similar to that employed elsewhere. Clinical ethics consultants, either individuals or small CEC subgroups, must begin by establishing the facts of the case. This involves obtaining information about several aspects: the patient's clinical background, medical indications, and prognosis (in terms of survival and quality of life). This can be done by reviewing the patient's clinical records, but sometimes it is necessary to clarify these aspects directly with the treating physician because, even if he/she did not request the referral, he/she should be aware of it. After collecting and clarifying this medical information, a meeting should be scheduled, ideally involving the patient, his/her family, and/or his/her representative in order to discover the patient's fears, expectations, beliefs, values, and preferences while also determining whether he/she has presented any advance directives and establishing the context of the patient and his/her disease. These efforts can be regarded as a prerequisite for any further steps, since they will make it possible to start a conversation aimed at attaining the best outcome possible for the patient.

It is crucial that both the professional who requested the referral and the treating physician participate in the meeting, since one of its objectives is to allow all those who play a role in treatment and care decisions to enrich the deliberative process. This is a meeting aimed at clarifying any possible doubts, after which the participants must seek the best possible solutions. In these meetings, the consultant works as a facilitator and sometimes as a mediator between different positions, encouraging the expression of values and paying attention to all the parties involved in the case (Fanning et al. 2015).

After collecting all the necessary information, and in order to reduce uncertainty regarding the case, the consultant—together with all those present in the meeting—analyzes the possible courses of action and finally offers recommendations. These recommendations consist in courses of action—complemented by their supporting ethical arguments—that the consultant notes on the patient's clinical record (Beca 2012). Thus, the consultation process informs the decision-making process, but does not replace those who must actually make these decisions, as the treating physicians are still ultimately responsible for their own choices.

Finally, two points must be considered: not all consultations result in recommendations; however, even if they do not, they still offer an opportunity to ponder and examine the ethical aspects of the situation in detail. The main objective of the consultation is not to yield recommendations, but to generate a process of interaction and deliberation during which the voices of all those involved are considered. Relevant to this issue, Joseph Carrese and the members of the American Society for Bioethics and Humanities recommend not assuming that a written report resulting from the consultation process will be understood without a verbal communication. It is worth reviewing and discussing the key aspects of these recommendations with the clinical team. This is also educationally relevant, since it can improve communication and offer learning opportunities (Carrese et al. 2012, Fanning et al. 2015).

Researchers have pondered whether the consultant must necessarily be a medical doctor or whether another professional could fulfill this role. In this regard, the literature suggests that, irrespective of his/her profession or specialization, the consultant must be a professional who is recognized by his/her peers, who has received adequate training in bioethics, who possesses certain personal features and competences that qualify him/her to occupy this position, and who is capable of understanding diagnoses, prognoses, clinical judgments, and treatment and care options.

The American Society for Bioethics Consultation has identified certain personal qualities and competences that a clinical ethics consultant must possess (Tarzian 2013). Before taking on their position, consultants must receive formal and tutorial training, for which only a handful of options are available. Also, it is worth noting that the consultant must have access to the patient's clinical file, where he/she will keep a record of the consultation process and its resulting recommendations; therefore, it would be complex and inadvisable for the ethical consultant to be external to the health care center where the patient is being treated.

New roles of CECs and clinical ethics consultation systems

Throughout this chapter, we have referenced some challenges that clinical ethics committees must face. In this section, we propose recommendations for CECs and clinical ethics consultation services or systems to address the challenges and new roles that they must navigate to deliver better service to patients. In our view, "better" service can only be delivered by determining the greatest "good" for each patient, considering his/her values, preferences, hopes, and context, which includes his/her loved ones and friends. This perspective has been disregarded or scarcely taken into account in contemporary medicine, given its scientific and technical focus and its strong dependence on management and funding. Nevertheless, it is a necessary ethical perspective to ensure that each patient can receive "good" health care, leading to improvement, rehabilitation, or a better end-of-life experience. This ethical consideration is necessary to achieve a patient-centered medicine, and therefore to achieve the humanization of medicine.

The following proposal is based on the authors' experience in clinical ethics committees and in clinical ethics consultation in Chile. This reflection has been enriched by sharing experiences with specialists from other hospitals and in other countries. Thus, these concepts are offered as ways to improve how clinical bioethics functions and how it can contribute to patients, our coworkers, those whom we teach about this, and the institutions where we are employed. These are not prescriptions or concrete projects; rather, they are ideas to be considered which readers can work to implement in specific health care centers taking into account their own characteristics and concrete possibilities.

The consultative role of clinical ethics committees

It is their core function, even though it can be insufficiently fulfilled for a variety of reasons. For example, it can be difficult to access the committee; in addition, institutional recognition of committees can also be limited. Cases are often belatedly presented to CECs, those who present them are almost exclusively physicians, with minimal involvement from other professionals, and presentations are largely technical-scientific in nature, offering little ethical or contextual information. Evidently, for a consultation to generate suitable responses, it must pose clear questions; furthermore, these questions must represent the doubts of all the professionals involved as well as those of the patient, his/her family, or his/her representative(s). Also, the facts to be analyzed must be presented in a way that is clear and understandable to all the members of the committee, who may not be specialists in the disease or treatment involved in each case. Consequently, it is a major challenge to establish who presents each case and what information he/she/they must provide, which means that the person requesting the committee's opinion must be aware of the patient's views, fears, desires, expectations, hopes, resources, and support networks. Case presentations with a clinical focus, but lacking clear information about the context and available resources, often cause the analysis process to yield conclusions that leave out aspects relevant to the patient's short- and long-term benefit. To tackle this frequent issue, each CEC could develop instructions or guidelines specifying the form and content of the documentation required for presenting cases before the committee.

Since case analysis is the most common task of bioethics committees, it is assumed that they operate well, but this is not always the case. In most committees, case presentation is followed by the clarification of the clinical facts and the opinions of some members. Unfortunately, the latter often tend to be intuitive and weakly supported. All committee members, even those with no formal bioethics training, should be familiar with the case analysis sequence utilized: clarification of the facts and the question(s) posed, explicit presentation of the values and value conflicts involved, and—finally—analysis of possible courses of action and expression of personal opinions to issue a joint recommendation. However, recommendations are frequently made without offering any ethical arguments, leaving those who did not take part in the deliberation process ill-equipped to understand the reasons for the courses of action proposed.

Since this failing is sadly common, CECs must rise to the challenge of improving case analysis and guaranteeing the systematic and well-organized utilization of their chosen methodology. This will allow each CEC member to become acquainted with the methodology and integrate it into his/her reasoning and argumentation process. In this regard, James Drane asserted that "if a suitable method is followed when making evaluations of a moral nature, the most severe mistakes can nearly always be avoided and defensible decisions can be adopted" (1990:415). Thus, we propose that all CECs should hold periodic meetings to reinforce their chosen clinical ethics analysis methodology and that the reports issued include ethical support for the recommendations made.

The normative role of clinical ethics committees

It is often deemed to be secondary or complementary. However, this function has a basic and important role: helping to ensure that institutional decisions focus on strengthening treatment, care, and disease prevention. Therefore, the challenge for CECs is to shift from a passive attitude to a proactive role comprising two lines of action: (1) conducting an ethical analysis of the institutional norms or projects submitted to them and (2) informing institutional authorities of the need to generate norms to improve or perfect aspects of patient care that the cases analyzed show to be deficient. Such deficiencies may involve informed consent forms, advance directives, terminal patient care, indication of insufficiently proven treatments, and treatment limitation, among other aspects. To issue normative proposals, CECs should work alongside the specialists or authorities in charge of each of these topics.

The educational role of clinical ethics committees

It is generally overlooked or may be obscured by their many other tasks, but it is enormously important. First, it is necessary to facilitate and demand that each committee member be suitably trained in bioethics, which requires the institution to fund and help members to enroll in bioethics courses, diploma programs, or Master's degree programs.

Committees should be periodically renewed by replacing members who have been unable to take part in specialization programs within a set period of two or three years. This measure can be said to be no different from the justified renewal of professionals who do not stay up-to-date in their discipline, only following a routine based on their experience and personal intuition. This lack of continuing education is common in CECs and has led to routine functioning, with case analyses yielding recommendations that are chiefly intuitive, lack specialized support, and fail to have an impact on the creation of criteria or the knowledge of those who present their doubts to the committee. Overall, few CECs have established that their members need training and continuing education, which they can pursue with suitable institutional support and commitment.

The educational role of CECs is logically not limited to its members; rather, it should benefit the health care institution as a whole. All the professionals and employees of an institution that treats and cares for patients require a basic level of ethical awareness or sensitivity that will lead them to be understanding of each individual user, patient, or family member. Some may have acquired this sensitivity due to their life experiences or temperament, but all of them need more information and training to cultivate this important competence. For this reason, CECs have responsibilities and challenges that, because they are rarely addressed, must be tackled by generating attractive educational initiatives aimed at each level of the organization: physicians, health care professionals other than doctors, auxiliary staff, administrators, and managers. It is necessary for all staff members to be well aware of the fact that all users or patients want to and must be treated with scientific, technical, and ethical excellence, and that clinical ethics committees are institutional bodies capable of taking on this educational role.

CECs can fulfill this role in a variety of ways, for instance, through the participation of CEC members in the meetings and courses of each work group, the general dissemination of basic ethical concepts, encouragement and invitations to attend bioethics courses, intra-institutional activities for the dissemination of clinical ethics experiences and the generation of educational "capsule courses" for new employees and for each work group, among other measures. For instance, these bioethics education initiatives may consist in clinical meetings focused on ethical deliberations about cases aimed at a variety of specialist groups, incorporating all the professionals who work at the health care center.

Conclusions

Research shows that clinical ethics consultation systems have a positive impact on decision-making processes in highly complex cases while also making valuable contributions to patients, family members, and clinical teams. Nevertheless, their development has been limited and varied in terms of modes of operation. The challenge for clinical ethics consultation systems is to cement their place within their health care institution, train professionals capable of becoming clinical ethics consultants. They also have to specify their mode of operation including how to request a consultation, who can do so, and when, establish how the system will operate in each case, regulate the participation of professionals and family members. It is necessary to determine methods to inform patients and relatives, describe how to keep records in the patient's clinical file, how to follow-up and support the parties involved in each case, analyze the implementation and mode of operation of the system, and study and disseminate the experience, among other tasks. Thus, the challenge for health care centers is to develop clinical ethics consultation systems suited to their institutional features, exploring their feasibility and the potential involvement of external consultants.

These new roles or challenges for clinical ethics committees and clinical ethics consultation systems entail substantial changes, as a complementary and often volunteer-based approach to bioethics must be replaced by a professional perspective adopted in response to the urgent needs of patients and institutions. To succeed in this endeavor, adequately specialized, trained, and paid professionals are required.

Despite the importance and richness of the deliberations that take place within CECs, having ethical consultation systems in place allows this resource to be regarded as a less distant and more useful tool for resolving ethical problems, thus preventing belated consultations. Like other medical specializations, bioethics must evolve to become a necessary activity, capable of offering serious responses to the need to move toward a medicine focused on seeking the greatest good for each individual patient. To offer a comparative example, only a few decades ago infections were treated by general practitioners or internists supported by laboratory cultures and antibiograms. However, infectiology has become professionalized to such a degree that nowadays specialists are needed to select treatments for patients with complex infections, since they must take into account immunological responses and a variety of interventions (not only of a pharmacological nature). Likewise, the severe and complex diseases affecting many patients around the world require the contribution of specialists in clinical bioethics who have been suitably trained to offer personalized answers.

These answers or recommendations must be professionally formulated, considering the clinical facts and the context of each case, as well as the values and preferences of each patient. This is the step that we refer to as the *professionalization of bioethics*, which should lead clinical ethics consultants to be seen not as "experts who define behaviors as either right or wrong" (Beca 2012:260), but as professionals who facilitate deliberation and decision-making in collaboration with the patient, his/her family, and the clinical team.

References cited

Abel, F. 2006. Comités de ética asistencial. Anales del sistema sanitario de Navarra 29: 75–83.

Agich, G.J. 2005. What kind of doing is clinical ethics? Theor. Med. Bioeth. 26: 7–24.

Altisent, R., T. Fernández-Letamendi and M.T. Delgado-Marroquín. 2019. Una nueva vitalidad para el futuro de los comités de Ética asistencial. Folia Humanística 13: 19–33.

Beca, J.P. 2012. Los consultores ético-clínicos. pp. 253–263. *In*: Beca, J.P., C. Astete and S. Carvajal (eds.). Bioética Clínica. Mediterráneo, Santiago de Chile, Chile.

Bravo, M. 2012. Comités de ética asistenciales. pp. 244–252. *In*: Beca, J.P., C. Astete and S. Carvajal (eds.). Bioética Clínica. Mediterráneo, Santiago de Chile, Chile.

Carrese, J.A., A.H. Antommaria, K.A. Berkowitz, J. Berger, J. Carrese, B.H. Childs and L. Wocial. 2012. HCEC pearls and pitfalls: suggested do's and don't's for healthcare ethics consultants. J. Clin. Ethics 23: 234–240.

Colby, W.H. 2019. Nancy Cruzan and the withhold versus withdraw dilemma. Am. J. Bioeth. 19: 1–2.

Decree 62. Aprueba Reglamento para la constitución y funcionamiento de Comités de Ética Asistencial, 2012. Santiago de Chile, Chile.

Drane, J.F. 1990. Métodos de ética clínica. Boletín de la Oficina Sanitaria Panamericana 108: 5–6.

Fanning, J.B., N.A. Garrison and L.R. Churchill. 2015. Beyond the recommendation: discerning achievable goals in clinical ethics consultation. Am. J. Bioeth. 15: 42–44.

Fernández Letamendi, T. 2016. Modelo de evaluación de la actividad de los comités de ética asistencial en Aragón. Ph.D. Thesis, University of Zaragoza, Zaragoza, Spain.

Ferrer, J.J. 2003. Historia y fundamentos en los comités de ética. pp. 17–42. *In*: Martínez, J.L. (ed.). Comités de bioética. Universidad Pontificia Comillas, Madrid, Spain.

Fox, E., M. Danis, A.J. Tarzian and C.C. Duke. 2022. Ethics consultation in US hospitals: a national follow-up study. Am. J. Bioeth. 22: 5–18.

Gómez, M.I. 2015. Los Comités de Ética Asistencial. pp. 170–179. *In*: Academia Chilena de Medicina (ed.). Reflexiones sobre Bioética: Seminarios de la Academia Chilena de Medicina (2011–2013). Academia Chilena de Medicina, Santiago de Chile, Chile.

González-Bermejo, D., M.D. Solano, J. Polache, A. Mulet and D. Barreda. 2020. Los Comités de Ética Asistencial y los Comités de Ética de la Investigación en España: organización, regulación y funciones. Revista de la OFIL. 30: 206–211.

Law 20.584. Regula los derechos y deberes que tienen las personas en relación con acciones vinculadas a su atención en salud, 2012. Santiago de Chile, Chile.

Martínez, K. 2007. Los documentos de voluntades anticipadas. Anales del Sistema Sanitario de Navarra 30: 87–102.

Morlans, M. 2021. Los Comités de Ética Asistencial. Fundació Grifols, Barcelona.

Myers, D. and J. Lantos. 2014. The slow, steady development of pediatric ethics committees, 1975–2013. Pediatr Rev. 35: 15–19.

Oficina de Bioética MINSAL-CEAM. 2019. Resultados de Encuesta a Comités de Ética Asistencial Sistema Público de Salud. Jornada de Bioética del Ministerio de Salud de Chile, Chile.

Portales, M.B. and J.P. Beca. 2021. Percepción de médicos intensivistas de adultos sobre el aporte de la consultoría ético-clínica. Revista médica de Chile 149: 997–1003.

Portales, M.B. 2022. Comités de Ética Asistencial y consultoría ético-clínica. pp. 145–152. *In*: Beca, J.P., C. Astete and S. Carvajal (eds.). Bioética Clínica. Mediterráneo, Santiago de Chile, Chile.

Pose, C. 2016. Los inicios de la consultoría ética: los comités de ética y su constitución. EIDON 45: 29–63.

Pose, C. 2017. La consultoría de ética clínica en la actualidad: revisión crítica de los modelos de mediación y propuesta de un modelo deliberativo. EIDON 48: 70–126.

Pose, C. 2020. La bioética, 50 años más tarde. EIDON 54: 11–23.

Post L.F. and J. Blustein. 2021. Ethical foundations of clinical practice. pp. 3–16. *In*: Post, L.F. and J. Blustein (eds.). Handbook for Healthcare Ethics Committees. Johns Hopkins University Press, Baltimore, USA.

Tarzian, A.J., ASBH Core Competencies Update Task Force. 2013. Health care ethics consultation: an update on core competencies and emerging standards from the American Society For Bioethics and Humanities' core competencies update task force. Am. J. Bioeth. 13: 3–13.

Velasco-Sanz, T. 2015. Origen de los comités de ética: Karen Ann Quinlan. pp. 159–168. *In*: Herreros Ruiz-Valdepeñas, B. and F. Bandrés Moya (eds.). Historia Ilustrada de la Bioética. Longares, Madrid, Spain.

Chapter 10
Critical Bioethical Approach to Health Crisis Scenarios

Eduardo Diaz Amado

Introduction: Bioethics, a cultural phenomenon of our time

Bioethics entered the intellectual, cultural, and academic mainstream when Potter (1971), an oncologist and biochemist professor at the University of Wisconsin, published *Bioethics. Bridge to the Future*. Bioethics synthesizes many of the fears and hopes of our time. Potter's view of bioethics in terms of a "bridge" was an invitation to protect life as well as the planet for future generations. Inspired by A. Leopold's work, Potter proposed a new ethics to guide our relations with nature and other living beings. He realized that life had become valuable only in terms of its utility, and animals, plants, and we ourselves as human beings were subjected to processes of commodification and reduced to simple objects of exploitation and research. In this grim scenario, Potter urged us to review and modify our current ideas about progress and development. But bioethics is a "bridge" in another sense, too. Potter considered a new relationship between humanities and social sciences, particularly ethics, on the one hand, and the biological sciences and medicine, on the other. Complex problems, like those of bioethics, required an interdisciplinary and transdisciplinary approach.

To some extent, bioethics is a kind of modern awareness at three levels. First, it asserts the intrinsic value of all forms of life and the moral duty of protecting them. In our time the weak equilibrium of life has been broken, with serious consequences. In his moving and prophetic letter

Instituto de Bioética, Pontificia Universidad Javeriana, Colombia.
Email: eduardo.diaz@javeriana.edu.co

to the President of the United States, in 1854, Chief Seattle had already warned us about the catastrophic consequences of the careless attitude of modern human beings to nature and to our brothers, the animals (Crowley 1999). Thus, bioethics, in a Potterian sense, seeks *to correct the course of the ship*, that is, our idea of progress and civilization that is destroying life. All living beings belong to the same family and the earth is our *common home*, like said Pope Francis' encyclical, *Laudato Sì* (Francis 2015).

Second, it reveals the Janus-faced functioning of modern science and technology, particularly in the biomedical field. Increasing our knowledge of nature, improving our healing abilities, and making industrial processes more efficient might be the happy face. But history has shown that abuses and deviations can happen, too. This is the sad face. Unfortunately, biomedical research has sometimes served unethical and questionable political agendas.

Third, it refers to the ethical and political transformation of medical practice and healthcare in the contemporary world. This transformation is intertwined with socio-political, economic, and cultural changes. Traditional ways of understanding the doctor-patient relationship went into a crisis. New actors, roles, and values emerged and changed such a relationship. Healthcare is nowadays provided by large health corporations and is not in doctors' hands anymore. In this situation, fighting domination and abusive behaviors became an important political aim of our day, as well as guaranteeing patients' rights to make choices (autonomy) and to get appropriate medical treatment (justice) (Diaz Amado 2020a).

In the process of becoming an independent and worthy discipline on its own, bioethics adopted the language of principlism or the *Georgetown Mantra* (Fox and Swazey 2008), which consists of four *prima facie* principles to analyze and solve ethical dilemmas in biomedical practice and research. These principles were laid out in the *Belmont Report* (1978) and in the T. Beauchamp and J. Childress' book, *Principles of Biomedical Ethics* (1979). Respect for autonomy, beneficence, non-maleficence, and justice is now the *lingua franca* of bioethics, which was easily appropriated by health professionals and society to talk about ethical problems arising in their field.

The simplicity of the four principles model for non-philosophers or expert ethicists explains why in healthcare and medical practice everyone has learned this language. Unfortunately, with the *bioethization* of these scenarios, medical ethics was deprofessionalized (Diaz Amado 2013). Regrettably, the bioethical lens is not powerful enough to examine all the ethical complexities of contemporary medicine, which have been reduced to a series of standardized ethical dilemmas.

Bioethics has also facilitated the assembling of new practices and subjectivities as well as novel methods and perspectives. The creation of national commissions, the arrival of bioethics committees (clinical ethics committees—the so-called Institutional Review Boards or IRBs), and the emergence of the *bioethicist* as an expert dealing with ethical dilemmas in healthcare and biomedical research reflect a new configuration of the ethical in biomedicine (Diaz Amado 2013). Bioethics seeks to promote a pluralist and inclusive point of view and, at the same time, guarantee respect for communities' and individuals' rights. This ethical commitment goes hand in hand with a new epistemological approach: bioethics is an interdisciplinary/transdisciplinary, deliberative, consensus-seeker, secular, and pluralist endeavor.

Bioethics: criticisms and its relationship to social sciences

The canonical definition in the W. Reich's *Encyclopedia of Bioethics* defines the new discipline as "the systematic study of the moral dimensions—including moral vision, decisions, conduct, and policies, of the life sciences and health care, employing a variety of ethical methodologies in an interdisciplinary setting" (2013:83). This definition obscured Potter's vision of a wide and global discipline devoted to reflecting on the meaning and protection of life. In addition, the preponderance of biomedical sciences and healthcare as the target object of bioethics was reinforced. Medical and biotechnological advances interrogated doctors and society about problems they were not prepared to deal with.

This model of bioethics, which is mainly focused on biomedicine and is based on the four principles, has been highly criticized. Some philosophers pointed to the unsatisfactory consistency of principlism. For instance, for Clouser and Gert (1990:220), "the principlism is mistaken about the nature of morality and is misleading as to the foundations of ethics" and Valdés (2011:70) refers to the "contradiction syndrome of principlism" or the contradiction among those principles when applied to particular cases. Other authors have warned about the excessive emphasis the American model of bioethics places on individuals' rights and its conceptualization of individuals as isolated from their socio-political and cultural contexts. A model that, moreover, has been conceived as universal; an assumption that has proved to be problematic.

Reich (2013), one of the American pioneers of bioethics, has drawn our attention to the loss of essential and primordial features such as the importance of approaching issues related to life sciences and humanities from a truly interdisciplinary, dialogical and inclusive perspective. Reich thinks that bioethics has been reduced to applied ethics and is now linked to a kind of aseptic argumentative exercise that, nevertheless, looks like a proper solution for the problems it was created to face. Yet, diverse cultural contexts and interesting methodological possibilities have not been appropriately taken into account and bioethics has remained fixed on ethical issues related to medical practice and healthcare. This situation has made bioethics boring, relegating its fundamental dimensions to the shadows (Reich 2013).

Elliot (1999) argues that a bioethics driven primarily by clinical worries enters a circularity by which the same concerns are discussed once and again, such as end-of-life issues, informed consent, and futility, to name but a few. The bioethics agenda is full of classical *ethical dilemmas* in medicine but remains short to examine in detail their structural, socio-political, economic, and cultural origins. Thus, it works as an auxiliary, expendable discipline in medical education and practice, which is supposed to provide quick solutions to several ethical problems making healthcare's bureaucratic machinery function (Diaz Amado 2013). For these reasons, Elliot (1999) believes that bioethics misses a good opportunity to philosophically enhance the techno-scientific dimension of medical practice and to carry out a humanistic reflection on life, death, and disease.

When social sciences took bioethics as an object of study, the taken-for-granted explanations of its origins and foundations underwent examination. Some of the issues examined included why bioethics is what it is and why it developed in the way it did. To be sure, the ethical dilemmas which bioethicists talk about do not appear spontaneously or in a vacuum; rather they emerge in a particular historical, socio-political, cultural, and economic context. Bioethics is not simply a speculative discipline; it is a specific kind of social action in a particular scenario that involves "the collision of social worlds, power differences, and the social construction of realities" (DeVries and Subedi 1998:xiv). It would be necessary, then, to ask why some issues become ethical problems in bioethicists' hands while others do not. On a deeper level, and taking into account how bioethics works, it is also compulsory to consider that "the clinical bent of bioethicists […] leads them to overlook some of the most profound ethical problems of medicine" (DeVries and Subedi 1998:xv).

Bioethics has been so successful in colonizing the biomedical scenario because it is functional to the way contemporary medicine and healthcare institutions work. Both the language of principlism and the theoretical framework of the Anglo-American analytic philosophy offer a palatable rhetorical interplay of arguments. However, mainstream bioethics lacks enough capacity to truly transform unjust and exploitative realities that are shaped by powerful economic, political, and cultural forces. Bioethics might rightly be seen as part of the contemporary reconfiguration of power in the biomedical field. Medical practice, biomedical research, and healthcare are particular territories whose population *must be governed* (Diaz Amado 2013).

Medicine is not in doctors' hands anymore and patients hardly decide on how healthcare works. Rather, medical practice and healthcare are under the control of large corporations and pharmaceuticals. These groups impose their own interests and modes of being (subjectivities). It is in this environment where bioethics emerges, challenging medical authority (Rothman 1991) and promoting the idea of free choice and respect for autonomy. In other words, bioethics is part of contemporary strategies to make medicine and healthcare manageable in accordance with the neoliberal episteme that has pervaded the whole social life (Diaz Amado 2013).

Behind the rise of bioethics is the assumption that scientists and doctors are ill-prepared to deal with ethical problems and that their training is not enough for them to make good decisions. For Wilson (2014), following R. Baker, D. Porter and R. Porter, the emergence of bioethics cannot be explained just "by simply framing it as a response to the moral dilemmas raised by new medical technologies during the 1960s" (p. 5). Philosophical and ethical problems linked to medical practice and its progress are not new in the history of medicine.

Then, it is urgent for bioethics to give room for "a more subtle analysis of the meaning of data about cultural preferences, political structures, and clinical situations" (Light and McGee 1998:5). The work and role of bioethics cannot be reduced to applying certain ethical theories or the four principles to the issues and problems populating the field. It is necessary to acknowledge that bioethical issues and problems are socially, politically, economically, culturally, and religiously shaped. In this way, it is of great value for bioethics to make visible the connections between norms, institutions, problems, and subjectivities (Light and McGee 1998).

Bioethics should not only be a philosophical and speculative enterprise; it has to grasp reality and transform the current state of things, particularly in the biomedicine/society relationships, promoting better scenarios for the future of humankind and the planet. This explains why it is fundamental to use empirical methodologies to support bioethical claims. Empirical bioethics began to gain relevance at the end of the 1990s. Empirical data must inform bioethics, which includes regulations, norms, strategies, prohibitions, and proceedings related to bioethical issues. Bioethics should be based on evidence. The empirical data illuminate and complement, not replace, moral reflection and deliberation in bioethics. It contributes to making visible those realities behind philosophical and speculative analyses. Social, natural, and biomedical sciences are part of the bioethical enterprise and deserve a place in the bioethical reflection as much as philosophy. Bioethics is an interdisciplinary/multidisciplinary field in which different professions and disciplines, along with their own methods, are welcomed, in order to carry out a wide and inclusive dialogue, to better understand problems, and to suggest possible solutions (Holm and Irving 2004).

Latin American bioethics, some contributions and inputs

Many bioethicists in Latin America would agree with Reich (2013) when suggesting that it is necessary "to renew the underpinnings, relationships, and activities of bioethics itself [and to counterbalance] the tendency towards reductionism in defining the scope and determining the methods of bioethics" (p. 82). This idea has motivated Latin American bioethicists to explore new horizons for the discipline. Thus, many of them revised bioethics ethical and epistemological foundations. Their aims were to define bioethics' scope beyond applied ethics and to expand the analysis of ethical dilemmas outside the field of sophisticated biomedical problems. In other words, Latin American bioethicists have sought to overcome a reductionist conception of bioethics by developing a profound commitment to important values linked to its original essence: ethical pluralism and an inter/transdisciplinary approach in its work (Tealdi 2008, Garrafa and Erig Osório de Azambuja 2009).

The Latin American bioethics view emphasizes the necessity for overcoming the dominant American model that is focused on the isolated, decontextualized, and dehistorialized individual. Autonomy has been wrongly assumed as an immanent, taken-for-granted property of human beings. But as several authors have shown, autonomy is a capacity that depends on well-established relationships in the community. Human beings live and interact in a complex net of relationships and mutual influences, among individuals themselves and between individuals and institutions. Justice also occupies a central place in the Latin American bioethics. It is not simply another principle of the principialist model; justice is rather a fundamental moral beacon in a scenario of profound socio-economic unbalances and power asymmetries (Garrafa and Erig Osório de Azambuja 2009).

In the field of biomedical research in the last few decades, there has been a prominent and progressive global regulatory development, including specific ethical principles and review instances (ethics committees). Still, millions of vulnerable people remain unprotected from abuses, exploitation, and deception. In the case of Latin America, this is a particular situation demanding special safeguards (Homedes and Ugalde 2012, 2016). The socio-political, economic, and cultural conditions explain why in Latin America there is much to do to protect the vulnerable and guarantee basic rights to all the participants in biomedical research projects, including clinical trials. Factors like poverty, weak regulatory institutions, lack of an adequate legal framework, and communities' ignorance of norms and rules worsen this situation.

It is necessary to work hard in order to prevent the repetition of painful, degrading, and humiliating examples of the past linked to biomedical research. For instance, the horrible experiments carried out by the Nazis, the infamous Tuskegee experiment, the double standard scandal with antiretroviral trials for HIV in Africa (Garrafa and Lorenzo 2008), and more recently, the non-compliance with ethical standards in the anti-Covid-19 vaccines research (McCarthy and McCarthy 2022). Biomedical research needs to be both scientifically excellent and ethically legitimate.

Many countries in Latin America have faced difficulties arising from double-standard practices, conflicts of interest, poor study design, the inability of the ethics committees to see the whole picture, and the irrelevance of many research projects for the communities in which they are carried out. Neither ethics committees nor well-intentioned declarations are enough to guarantee truly ethical research (Garrafa and Lorenzo 2008, Garrafa 2016). To face this situation, different strategies and answers have been proposed. Some authors have suggested adopting the *Universal Declaration on Bioethics and Human Rights* (DBHR) instead of the *Declaration of Helsinki*, which has been hardly criticized for opening the door to the legitimization of the double standard in clinical trials (Garrafa et al. 2010, Tovar 2020).

The relevance of the DBHR rests on the fact that it was signed by States. Its content and structure reflect a particular commitment to the necessities and aspirations of the peripheral countries. In the DBHR, bioethics and human rights are integrated, signaling a necessary collaborative and inspiring interaction (Albuquerque and Paranhos 2020). The DBHR enhances the ethical perspective of bioethics by going beyond the well-known four principles. Principles like human dignity, respect for human vulnerability and personal integrity, and the obligation to share benefits are just some of the contributions of this declaration to the ethical foundations of bioethics.

For Latin America, bioethics is not only an attempt to understand and discuss issues related to biological sciences and medicine but also an opportunity to transform its own reality. Beyond the analytical tone and the normative trend of the dominant model of bioethics, it is necessary to underline the emancipatory vocation of its beginnings. Bioethics is equipped with powerful theoretical and practical elements to analyze the reality, particularly when biomedical sciences are implicated, and to orientate human action, especially when it involves managing or transforming life. As Tealdi (2008) has argued, bioethics is interested not only in the techno-scientific developments but also in the human interaction to face these problems.

The 6° World Congress of Bioethics, which took place in Brasilia in 2002, was a turning point for the global academic bioethics establishment. In this event, a new comprehension of what bioethics is for was born. Supporters of Latin American bioethics proposed the so-called *intervention bioethics,* which distinguishes between persistent and emergent problems (Garrafa 2008). The former refers to those problems related to socioeconomic inequities, democratic failures, and power asymmetries, which include, for instance, poor medical attention, bad living conditions, and environmental degradation. The latter has to do with those issues that people and mass media usually relate to bioethics, that is, ethical dilemmas linked to high-tech biomedical developments as well as medical ethics problems in the context of the bureaucratic and commodified healthcare services of our time.

Intervention bioethics seeks better answers and appropriate criteria for the analysis of large problems and collective conflicts related to persistent bioethical dilemmas that are common in poor and developing countries (Garrafa 2008). Intervention bioethics is deeply committed to equity and acknowledges socio-economic differences and basic unmet needs among the most vulnerable groups. For this reason, this bioethics model born in Latin America seeks to implement social justice, based on human rights and citizenship. Intervention bioethics opposes the moral imperialism and neocolonialism that, hidden in the dominant principialist model, pervades the whole scenario of biomedical practice and research. Moral imperialism, in bioethics, "is the aim of imposing, through different forms of coercion, moral standards from specific cultures, geopolitical regions, and countries on other cultures, regions, or countries" (Garrafa and Lorenzo 2008:2220). Moral imperialism includes also academic alliances, courses, and partnerships through which mentalities and opinions are aligned with the dominant bioethics model (Garrafa and Lorenzo 2008).

Protection bioethics has also been promoted in Latin America, for example, by Kottow (2008), to counterbalance the negative effects of exploitative power dynamics in the contemporary world. Power relations are not just a theoretical construct; they are forces determining people's lives, either by giving them opportunities and satisfaction or rather by making them miserable and unhappy. The weakening of the State as a result of neoliberal policies has left people and communities adrift and without adequate protection. *Protection bioethics* invites us to seriously take into account both the needs and the suffering of people "for whom there is no consolation in philosophy but rather in receiving assistance" (Kottow 2008:167).

Bioethics, biopower, and the Covid-19 pandemic

On February 12, 2020, the outbreak of the then-unknown disease was named: *Coronavirus Disease 2019* or *Covid-19*. This epidemic that only affected China at the beginning quickly became a pandemic whose severity was initially underestimated by the leaders of many countries, making its treatment and management more complicated. In Latin America, some presidents denied the problem for months, including Mexico and Brazil. The idea of avoiding panic and adverse economic repercussions led some governments to react with unclear and ambiguous answers. But once the pandemic showed its destructive power and hospitals started to be overcrowded with a rapid increase in death numbers, the importance of *experts'* opinions rocketed (Lavazza and Farina 2020). As the disease rapidly evolved into an apocalyptic-like situation, exponentially affecting more and more people, politicians and decision-makers turned their attention to healthcare professionals, particularly intensive care specialists and epidemiologists (Litewka and Heitman 2020).

In the beginning, fear and terror expanded throughout society. Mass media (TV, radio, newspapers) and social networks played an important role in the management of the pandemic, amplifying emotions and creating a reality-like spectacle (Diaz Amado 2020b). When confinement was implemented, life started to be lived on the screens. Face masks became compulsory and restrictions to continue with normal social life were rapidly accepted. It looked like a dystopic situation, but it was real. People's willingness to live in this situation can be relatively easy to explain: we all were already living on the screens before the pandemic, and we had already learned to hide our faces from each other in today's competitive and hypocritical societies, so wearing face-masks was not a terrible problem in the end (Diaz Amado 2020b).

According to Lavazza and Farina (2020), we trust experts because their advice has produced important and successful practical effects. From the decision-making process perspective, resorting to experts "is a way to curb any potential controversy". They are supposed to possess "objective knowledge" and to usually act rationally. This explains why epidemiologists became the *pop stars* of the Covid-19 spectacle, as this pandemic was, for the first time in history, globally shaped and lived on the screens. Daily conversations used to include quotes from one or another expert about what was going to happen next or how to prevent or fight the disease. But paradoxically, there was a significant level of distrust against experts. There were two main reasons to explain this, particularly among laypeople.

First, sometimes, experts' opinion reflects more their political, religious, or ideological views than a technical or reasonable perspective. Second, the way autonomy is assumed nowadays: people are convinced they can think or do whatever they want. In the current individualistic environment, somebody else's opinion is usually understood as undue interference. The consequence is that one's ignorance is reinforced instead of challenged. We should remember, in any case, that experts' advice is not "as technical and neutral as it is supposed to be" (Lavazza and Farina 2020:4).

Experts are related to power and the power perspective is helpful to comprehend many of the dynamics that emerged during the pandemic. It includes critically approaching history and how it is made, how subjectivities are shaped, how diverse social relations are intertwined, and how practices, discourses, and institutions influence social life. Furthermore, how people are governed, that is, what strategies are used to "conduct the conduct", as Foucault (1991) puts it. This refers to *governmentality*, a combination of rationality and governmental actions operating on people who see themselves as free agents (Dean 1999).

However, as Foucault (1991) describes it, in modern society, there are other kinds of power at work. *Sovereign power* is related to the privilege of the king to take life or let live. This power is repressive, as it is the force used by the State, for instance, through the army. *Discipline* is a kind of power that is exercised over individuals' bodies and behaviours through practices imposed by institutions. This power is present in schools and prisons as well as in medical education institutions and hospitals. In healthcare systems, both professionals and patients are subjected to modes of being and sophisticated surveillance strategies.

From a Foucauldian point of view, power is not an ontological characteristic of either subjects or objects; it reveals itself in terms of a relation or a strategy. Power can be repressive, but in Foucauldian terms, it is a productive force to create subjectivities and knowledge. It works in the whole social body. It is exercised not only by the State or institutions, but also by discourses and practices. And most importantly, it is present in the way individuals are governed and govern themselves in the name of freedom. Freedom is not just an option or a right; it is a government strategy in current neoliberal times (Diaz Amado 2013).

The Covid-19 pandemic revealed the importance of inquiring about how power functions and its effects in terms of biopolitics. The term biopolitics describes a particular kind of power that is exercised over individuals' bodies and populations. Inspired by Foucault's work, Sarasin (2020) identifies three models of thought about three infectious diseases in history, which are related to biopolitics. Each one represents three different forms of government and power exercise: leprosy, plague, and smallpox.

Power separates and excludes as exemplified by the way leprosy was managed in the past (*exclusion*). In contrast, the government model to manage plague consisted of *reclusion* (confinement) rather than *exclusion* and the compelling of subjects to obey certain rules (disciplinary practices). In this model, bodies should remain docile and productive. Finally, under the smallpox model, freedom is "inoculated". The neoliberal creed has convinced us that we live in a world where we are allowed to make as many choices as possible. Choices, not thoughts. Discipline and surveillance can be put behind; individuals are now controlling themselves. All these models are present simultaneously in current societies. However, the pandemic has been useful to unveil that "[w]here power wants to return from the smallpox model to the plague model, it becomes authoritarian and eventually totalitarian" (Sarasin 2020:6).

During the pandemic, entire populations were subjected to diverse forms of confinement, police control, chasing, and surveillance. This was done in the name of health and life, using legal mechanisms and the power of the State (sovereign power). Burris et al. (2021) refer to "legal epidemiology" or the form that epidemiological and public health interventions took on when adopting the same *modus operandi* of law: "laws, unlike pills, are administered at the group level and, [...] they produce different benefits and harms based on social position and other factors" (Burris et al. 2021:1974). Although repressive and surveillance strategies were used, it is interesting that the vast majority of people quickly assumed a subjection disposition. They were just obeying. According to Lorenzini (2021), it is not that surveillance and control mechanisms increased with the pandemic as we already were "docile, obedient, biopolitical subjects". This explains why it was so easy to implement restrictive measures during the pandemic.

In Latin America, however, for millions of people the real choice was not about contributing or not to slowing down the transmission of the virus by keeping themselves under lockdown. It was about how to survive. They had to go out and work despite risks and prohibitions. However, many of us just accepted all the restrictions and the loss of rights. For Lorenzini (2021), if we just insist on coercive measures, on being confined, controlled, and "trapped" at home during these *extraordinary* times, we risk overlooking the fact that disciplinary and biopolitical power mainly functions in an automatic, invisible, and perfectly ordinary way—and that it is most dangerous precisely when we do not notice it.

It was clear that people of the higher social strata had the possibilities and mechanisms to survive during the confinement regardless of whether or not they were forced to stay at home. Otherwise happened to millions who did not have such an option. The exclusion experienced by these people became worse during the pandemic since they did not have access to mechanisms and instruments of social protection. In Latin America, to make things worse, governments in general, were not efficient to provide the necessary support to the most vulnerable groups. Additionally, corruption played a major role. In the case of Colombia, for instance, social and academic organizations denounced a lack of transparency in the use of resources (millions of dollars) allocated to supposedly cope with the situation (El Espectador 2021).

Agamben (2017) had drawn our attention, even before the pandemic, to the fact that in contemporary societies *bare life* seems to be the one that matters; that is, human beings are becoming important in terms of biological entities. *Bios*, which is the Greek term to describe a life that is lived in society—the political and ethical life, has been devalued. Then, what kind of life matters to bioethics? Bishop (2009) has underlined that this is an important distinction for bioethical work. Referring to the case of Schiavo in 2005, Bishop argues that conservative bioethicists tend to stress the protection of bare life (*Zoè*), while liberal bioethicists privilege political life (*Bios*). What is the kind of life that matters to epidemiology?

Biopolitics means that human life has acquired a political significance in our time. Human beings are important not only because they are workers, but because they are living beings with a biological potential, for instance, as it is expressed in their genome or neurological functioning. This means that human beings "are no longer governed only, nor even primarily, as political subjects of law, but as living beings who, collectively, form a global mass—a 'population'—with birth and death rates, morbidity, average life expectancy, and so on" (Lorenzini 2021:41). This is exactly what happened during the Covid-19 pandemic. This explains why measures to avoid contagion were preeminent as a means for the protection of the "biological mass" that labor and produce. The protection of individuals as free subjects and citizens entitled to a bundle of legal rights was relegated to second place.

Bioethics, Covid-19 pandemic, and Latin America

The pandemic of Covid-19 has played the role of a negative film exposure in the dark environment of current societies. Soon after the outbreak in early 2020, the meanings of the first great pandemic of the 21st century were revealed. First, the infection caused by Sars-Cov-2 was not simply a biological event: a new lethal virus for human beings, particularly those with certain comorbidities and of advanced age. Rather it was a hard test for our social structures and way of living. It also meant a test in other ways: on the value of human life, the type of relationship we human beings have with nature, the ethical foundations of contemporary societies, the role of several institutions meant to promote health and life, and the indispensable balance that should exist between science, medicine, politics, and economy.

In Latin America, the Covid-19 pandemic unveiled serious problems: weak healthcare systems and poor biomedical research, including "a weak regulatory body, immature scientific community, and deficient scientific and ethical training in universities" (Fuentes Delgado and Angulo Bazán 2020:11). Errors and mistakes that led to failures to implement an adequate and on-time medical attention during the pandemic, as well as appropriate public health policies, were related to "ideological misconceptions and denial among Latin America's political leaders [that] prevented timely preparations for the pandemic and added to chronic governance problems" (Litewka and Heitman 2020:69). Mexican and Brazilian cases were paradigmatic examples in this sense. To make things worse, some governments went so far as to use the pandemic tragedy to obtain political profits (Litewka and Heitman 2020).

However, we cannot blame the pandemic for all the ills in Latin America. Before this exceptional time, problems to access high-quality medical attention and healthcare services were already present, reproducing deep socio-political and economic asymmetries: "Covid-19 has exposed the extent of such inequalities across the Latin America region" (Saldaña 2020:1434–1435). Of course, the pandemic worsened the precarious situation of hospitals and public health in Latin America, where scarcity and public distrust have been customary (Litewka and Heitman 2020).

Figures published by WHO in May 2022 showed that 54% of the total number of excess deaths during the two years of the pandemic occurred in low-middle and low-income countries (Grimley et al. 2022). Such number refers to "how many more people died than would normally be expected based on mortality in the same area before the pandemic hit" (Grimley et al. 2022). Within such countries, infection rates and deaths were higher among underprivileged groups. For instance, indigenous people in the Amazon region, and afro-descendants had a 62% higher risk of dying than white people in Brazil. Moreover, of the total number of healthcare workers, 70% were women and they were the most subjected to attacks from people who feared contagion (Saldaña 2020). Refugees and displaced people were also disproportionally affected. Then, in the face of this situation, bioethics must assume the voice of the voiceless and undertake the necessary political critique and analysis of the current state of things.

There are several situations requiring the participation of bioethics, although the first step is recognizing the biopolitical side of the pandemic. It is biopolitical because it includes police actions, confinement, face masks, citizens' detention, and borders closing. And the biopolitical analysis includes considering old and new forms of power exercise. Yet, before the Covid-19 pandemic, we were already living in a biopolitical time. In our time, life has been reduced to a manageable good according to the laws of the market (Rose 2007).

Bioethics' role is not exclusively finding out ethical principles to solve problematic situations related to biological sciences and medicine. Bioethics should include in its analyses the influence of different practices and institutions, structures and discourses, and subjectivities and apparatuses of truth (*dispositifs*) in the production of "bioethical issues". This includes a better understanding of the interface between biological sciences, medicine, and society. Bioethics must avoid reproducing the same inequities, asymmetries, and abuses it is denouncing. All of this means power analyses for bioethics (Diaz Amado 2013).

The pandemic scenario has proved the need of overcoming the autonomy-centered approach to bioethics and, instead, promoting an alternative model that emphasizes public health, collective good, and democracy (Garcia Alarcon and Montagner 2017, Saldaña 2020). Yet, the emphasis was put on the use of scales and algorithms in the decision-making process to solve the ethical dilemmas arising in the exceptional circumstances of the pandemic. In the Colombian case, for instance, bioethics scholars from different academic and professional backgrounds as well as the Ministry of Health and Social Protection outlined guidelines based on different ethical principles and theories (ANM 2020, Rueda et al. 2020).

Leaving all the decisions exclusively in doctors' hands was considered a bad idea, then special ethical committees and advisory teams were formed to help in the decision-making process. These guidelines included orientation on palliative care and psychological support for both patients and their families during the hardest time of the pandemic. Yet, these guidelines were not always the result of a wide and inclusive consensus. For instance, decisions about confinement or available treatment for certain groups like the elderly or disabled people are still a matter of strong discussion.

Like other places in the world, in Latin America, the best way to make ethical decisions and what principles should be considered was discussed. There were multiple debates on how to decide on the pre-eminence or futility of different practices during the decision-making process. For some experts, the aim of healthcare attention during the pandemic should have been just saving as many people as possible (Saldaña 2020). For others, it was necessary to take into account some crucial determinants or conditions like age, presence of comorbidities, and chances of survival and full recovery in the long and short term (prognosis). However, these approaches and conditions were not just in there. They resulted from different influences such as socio-economic asymmetries, racism, structural violence, and a lack of access to proper healthcare services. It is not mere biology or destiny.

According to Garcia Solis (2021), with the involvement of bioethics in the epidemic, we try to face the tensions and conflicts that arise from the clash between cultures, individual and collective interests. The protection of vulnerable groups must be highlighted. Health systems must be transformed and give impulse to initiatives that help ameliorate the exclusion, inequity, and barriers to get the integral health interventions promptly.

The tension between individual and collective rights was a sensitive point of the debate. Much of the literature on this tension was defined as a Shakespearian dilemma: saving lives or protecting the economy—*to be or not to be*. In the case of Latin America, was this really a dilemma, or was it a pseudo-dilemma? What if pre-pandemic economies were deadly enough for millions of people? Should these economies be saved or, rather, changed? Will we be able one day to really acknowledge that current economies kill as many people as the pandemic? The question today is whether implementing measures like, for example, making older people remain inside and vaccinating them first were the appropriate and more urgent measures.

In Colombia, for instance, there was an interesting expression of resistance against the restrictions imposed by the authorities during the pandemic. It was called the "grey hair revolution" (*la revolución de las canas*), as it was promoted by a group of influential elderly people (El Espectador 2022). Recently, in an interesting decision, the Constitutional Court of Colombia ruled that "the fundamental rights to freedom of movement, free development of personality, equality and decent work" of older people must be guaranteed and that restrictions imposed on them during the pandemic had violated those rights (El Espectador 2022).

Conclusions

In the current situation of global and sophisticated capitalism, unbalances between rich and poor countries tend to be perpetuated. This reality has so many important consequences that should not be overlooked by bioethics. Medical attention, healthcare services, biomedical research, natural resources policies, and measures to counterbalance the environmental crisis become bioethical issues that are approached as if they were "universal" in their genesis and solution. As a result, these issues and problems end up being seen and understood in an incomplete and biased way, while the structural, political, and socio-economic causes remain hidden or neglected. All of this constitutes the effects of power that ought to be analyzed in bioethics.

It is important to bear in mind that bioethics is not free from the influence of power imbalances and asymmetries. The dynamics of inequality and exploitation also alter the good intentions under the umbrella of bioethics. It is not rare that an academic field replicates certain socio-political and cultural distortions. Then, bioethics has to be seen as a scenario of confrontations, conflicts, and contradictions; a reflection of our own time. Perhaps it would be wise to approach bioethicists' work from different latitudes and cultures in the same way that a physician examines a patient, that is, using the clinical method, which is close to the method of ethics: connecting a particular story with diverse variables and seeking to find the meaning of what is happening. Bioethics is not only the strategy we have invented to solve our own social pathologies and prevent dreadful scenarios in the future, but it is also itself a symptom of our present and an expression of what we have become.

In the encounter of bioethics with the Latin American reality, I emphasize six important matters. First, the American, principlist and biomedical high-tech-centered bioethics is neither absolute nor universal, and then other bioethics are possible and necessary. Second, these other bioethics should consider particular contexts and realities as well as different cultures and values. Third, bioethics should avoid becoming an empty rhetorical exercise or mere philosophical speculation. Fourth, it is an unacceptable reduction to make bioethics only an ethical reflection on high-tech issues in biomedicine or a new kind of medical ethics. Fifth, the vocation and ability of bioethics to promote a wide discussion of several problems relevant to society, from an interdisciplinary/transdisciplinary, secular, plural, and inclusive perspective. And sixth, bioethics has the potential to change relationships and institutions.

Even if bioethics works mainly at the biomedical level, it requires taking into account the socio-political, economic, and cultural contexts if it wants to successfully achieve its goals. If "[e]pidemiology is the study of the distribution and determinants of disease and other health states in human populations by using up comparisons, to improve population health" (Broadbent 2013:17), bioethics must inform this endeavor so that we do not end up being managed just as biological beings or as human resources, rather our fundamental rights and dignity as human beings are really guaranteed and respected.

References cited

Albuquerque, A. and D. Paranhos. 2020. Direitos humanos como fundamento teórico-prático da Bioética de Intervenção. Rev. Redbioética/UNESCO 2: 14–24.

Agamben, G. 2017. Homo sacer. El poder soberano y la vida desnuda. Adriana Hidalgo Editorial, Buenos Aires.

ANM – Academia Nacional de Medicina de Colombia. 2020. Recomendaciones de la Academia Nacional de Medicina de Colombia para enfrentar los conflictos éticos secundarios a la crisis de Covid-19 en el inicio y mantenimiento de medidas de soporte vital avanzado.

Beauchamp, T.L. and J.F. Childress. 1979. Principles of Biomedical Ethics. Oxford University Press, New York.

Belmont Report. 1979. https://www.hhs.gov/ohrp/regulations-and-policy/belmont-report/index.html.

Bishop, J. 2009. Biopolitics, Terri Schiavo, and the Sovereign Subject of Death. J. Med. Philos. 33: 538–557.

Broadbent, A. 2013. Philosophy of Epidemiology. Palgrave-Macmillan, Basingstoke.

Burris, S., E. Anderson and A. Wagenaar. 2021. The "Legal Epidemiology" of pandemic control. N. Engl. J. Med. 384: 1973–1975.

Clouser, K. and B. Gert. 1990. A critique of principlism. J. Med. Philos. 15: 219–236.

Crowley, W. 1999. Chief Seattle's Speech. https://www.historylink.org/File/1427.

Dean, M. 1999. Governmentality: Power and Rule in Modern Society. SAGE, London.

DeVries, R. and J. Subedi. 1998. Preface. pp. xi–xix. *In*: DeVries, R. and J. Subedi (eds.). Bioethics and Society. Constructing the Ethical Enterprise. Prentice-Hall, Upper Saddle River, New Jersey, USA.

Diaz-Amado, E. 2013. The Transformation of The Medical Ethos and The Birth of Bioethics in Colombia. A Foucauldian Approach. Doctoral Thesis, Durham University. Durham, U.K.

Diaz Amado, E. 2020a. La relación clínica: una oportunidad para ser más humanos. pp. 155–176. *In*: Vásquez Ghersi, E. and M.A. Polo Santillán (eds.). Bioética. Una perspectiva desde América Latina. Universidad Antonio Ruiz de Montoya, Lima, Perú.

Diaz Amado, E. 2020b. Entre seguridad y libertad. pp. 38–57. *In*: Pérez Benavides, A.C. (coord.). Pensamientos virales: las ciencias sociales y humanas en tiempos de crisis. Pontificia Universidad Javeriana, Bogotá, Colombia.

Elliot, C. 1999. A Philosophical Disease: Bioethics, Culture, and Identity. Routledge, New York.

El Espectador. 2021 (18 July). Corrupción en tiempos de COVID-19: ¿Cuánto le cuesta a Colombia? https://www.elespectador.com/economia/corrupcion-en-tiempos-de-covid-19-cuanto-le-cuesta-a-colombia-podcast/.

El Espectador. 2022 (26 March). Corte Constitucional amparó los derechos de miembros de la "Revolución de las canas". https://www.elespectador.com/salud/corte-constitucional-amparo-los-derechos-de-participantes-de-la-revolucion-de-las-canas/.

Foucault, M. 1991. Governmentality. pp. 87–104. *In*: Burchell, B., C. Gordon and P. Miller (eds.). The Foucault Effect. Studies in Governmentality. The University of Chicago Press, Chicago, USA.

Fox, R. and J. Swazey. 2008. Observing Bioethics. Oxford University Press, New York.

Francis. 2015. Encyclical Letter Laudato Si' of the Holy Father Francis on care for our common home. https://www.vatican.va/content/francesco/en/encyclicals/documents/papa-francesco_20150524_enciclica-laudato-si.html.

Fuentes Delgado, D. and Y. Angulo Bazán. 2020. Desafíos bioéticos en el contexto de la pandemia por el COVID-19 en Latinoamérica. Revista Latinoamericana de Bioética 20: 11–13.

Garcia Alarcon, R. and M.A. Montagner. 2017. Epistemología de la bioética: extensión a partir de la perspectiva latinoamericana. Revista Latinoamericana de Bioética 17: 107–122.

Garrafa, V. 2008. Bioética de intervención. pp. 161–164. *In*: Tealdi, J.C. (dir.). Diccionario Latinoamericano de Bioética. Unesco-Red Latinoamericana y del Caribe de Bioética, Universidad Nacional de Colombia, Bogotá, Colombia.

Garrafa, V. and C. Lorenzo. 2008. Moral Imperialism and multi-centric clinical trials in peripheral countries. Cad. Saúde Pública 24: 2219–2226.

Garrafa, V. and L. Erig Osório de Azambuja. 2009. Epistemología de la bioética - enfoque latino-americano. Revista Colombiana de Bioética 4: 73–92.

Garrafa, V., J.H. Solbakk, S. Vidal and C. Lorenzo. 2010. Between the needy and the greedy: the quest for a just and fair ethics of clinical research. J. Med. Ethics 36: 500–504.

Garrafa, V. 2016. Una bioética latinoamericana. Interview. https://redbioetica.com.ar/una-bioetica-latinoamericana-volnei-garrafa/.
Grimley, N., J. Cornish and N. Stylianou. 2022 (5 May). Covid: World's true pandemic death toll nearly 15 million, says WHO. BBC News. https://www.bbc.com/news/health-61327778.
Holm, S. and L. Irving. 2004. Empirical research in bioethics: report for the european commission. pp. 131–155. *In*: Holm, S. and M.F. Jonas (eds.). Engaging the World: The Use of Empirical Research in Bioethics and the Regulation of Biotechnology. IOS Press, Amsterdam, Netherlands.
Homedes, N. and A. Ugalde (coords.). 2012. Ética y ensayos clínicos. Lugar Editorial, Buenos Aires.
Homedes, N. and A. Ugalde. 2016. Health and ethical consequences of outsourcing pivotal clinical trials to latin america: a cross-sectional, descriptive study. PLoS ONE 11: e0157756.
Kottow, M. 2008. Bioética de protección. pp. 161–164. *In*: Tealdi, J.C. (dir.). Diccionario Latinoamericano de Bioética. Unesco-Red Latinoamericana y del Caribe de Bioética, Universidad Nacional de Colombia, Bogotá, Colombia.
Lavazza, A. and M. Farina. 2020. The role of experts in the Covid-19 pandemic and the limits of their epistemic authority in democracy. Front. Public Health 8: 356.
Light, D.W. and G. McGee. 1998. On the social embeddedness of bioethics. pp. 1–15. *In*: DeVries, R. and J. Subedi (eds.). Bioethics and Society. Constructing the Ethical Enterprise. Prentice-Hall, Upper Saddle River, New Jersey, USA.
Litewka, S. and E. Heitman. 2020. Latin American healthcare systems in times of pandemic. Developing World Bioeth. 20: 69–73.
Lorenzini, D. 2021. Biopolitics in the time of coronavirus. Critical Inquiry 47: S40–S45.
McCarthy, M.S. and M.W. McCarthy. 2022. Ethical challenges of prospective clinical trials during the COVID-19 pandemic. Expert Review of Anti-Infective Therapy 20: 549–554.
Potter, V.R. 1971. Bioethics. Bridge to the Future. Prentice-Hall, New Jersey.
Reich, W. 2013. A corrective for bioethical malaise: revisiting the cultural influences that shaped the identity of bioethics. pp. 79–100. *In*: Garret, J., F. Jotterand and D.C. Ralston (eds.). The Development of Bioethics in the United States. Springer, New York, USA.
Rose, N. 2007. The Politics of Life Itself. Biomedicine, Power, and Subjectivity in the Twenty-First Century. Princeton University Press, New Jersey.
Rothman, D. 1991. Strangers at the Bedside. A History of How Law and Bioethics Transformed Medical Decision Making. Basic Books, New York.
Rueda, E.A., A. Caballero, D. Bernal, L. Torregrosa, E.M. Suárez, F.E. Gempeler and N. Badoui. 2020. Pautas éticas para la asignación de recursos sanitarios escasos en el marco de la pandemia por COVID-19 en Colombia. Revista Colombiana de Cirugía 35: 281–289.
Saldaña, A. 2020. Bioethical Guidelines of 'Extreme Triage' Under Covid: The question of 'Possible Lives' in Latin America. Bionatura. Latin American Journal of Biotechnology and Life Sciences 5: 1434–1437. https://revistabionatura.com/files/2020.05.04.27.pdf.
Sarasin, P. 2020 (March 31). Understanding the Coronavirus Pandemic with Foucault? *In*: G+C Blog. https://blog.genealogy-critique.net/essays/254/understanding-corona-with-foucault.
Tealdi, J.C. (dir.). 2008. Diccionario Latinoamericano de Bioética. Unesco-Red Latinoamericana y del Caribe de Bioética, Universidad Nacional de Colombia, Bogotá, Colombia.
Tovar, Y. 2020. Lessons of bioethics in legal education to face the challenges of COVID-19 in the protection of individuals and groups of special vulnerability: an approach from a Latin American perspective. Ann. Bioethics Clin. App. 3: 000143.
Valdes, E. 2011. The problem of principlism. Anamnesis 5: 53–80.

Chapter 11
The Challenge of Teaching Bioethics

Irene Cambra-Badii

Introduction

Teaching bioethics has always been a challenge

As seen in previous chapters in this book, bioethics is a relatively new a discipline. In the last two centuries, scientific discoveries and technological innovations in biomedical sciences have immensely improved the lives of most human beings. At the same time, however, these advances have posed new ethical problems in healthcare, making the teaching of bioethics even more complex than it already was. A universal answer remains elusive to the question of how best to address the challenges and bioethical concerns brought about by biotechnological changes in healthcare education.

One response has been a return to the humanities (Osler 1899, Miles 2009, Solomon 2015, Orefice et al. 2019). The humanities comprise the set of disciplines related to human beings, our knowledge, and our culture: art and art history, literature and comparative literature, history, theology and religion, and philosophy, among others.

As pointed out in Martha Nussbaum's *Not for profit: why democracy needs the humanities* (2016) and in Nuccio Ordine's *The usefulness of the useless* (2017), all students, not only those majoring in humanities, need to be exposed to humanistic disciplines to develop generic competencies like analytical and critical thinking, leadership, innovation, oral and written

Universitat de Vic – Universitat Central de Catalunya, Spain.
Email: irene.cambra@uvic.cat

communication, argumentation, time management, and social empathy. Nussbaum and Ordine's works show why the humanities are necessary, perhaps more than ever, in pre-university as well in university education in a world where the prioritization of profit over people can create serious confrontations.

Just as the humanistic disciplines are often relegated to humanities departments, bioethics is often taught only in health sciences education. Bioethics courses in health sciences often include the following topics: ethical theory, truthfulness, autonomy, informed consent, confidentiality, issues related to the beginning and the end of life, clinical relationships, human reproduction, genetics, vulnerability and patients' rights, distribution of health resources, and research.

Although bioethics is logically of great importance in degree programs focusing on patient care, bioethical dilemmas can appear in anyone's life, regardless of one's profession. Bioethical dilemmas do not only arise in medicine, and bioethics classes are certainly not the only context in which future healthcare professionals will have to deal with them. Bioethics transcends medicine, emphasizing the articulation between moral values and scientific-technical knowledge, underlining human beings' responsibilities to one another, to nature, to the future of humanity, and to the planet. Generalizing bioethics education to target all students from pre-school through high school and higher education is fundamental to teach a new generation of human beings to coexist with one another.

Despite the apparent consensus on the importance of teaching bioethics, in practice, very few courses deal with bioethics. One analysis found that only 2% of 4000 undergraduate courses offered at a large university in the United States had an ethics component (Beever et al. 2021, ten Have 2021). The inclusion of ethics components can mostly be attributed to the efforts of enthusiastic teachers, because institutional support and systematic efforts to create a future generation of ethics teachers are often lacking (ten Have and Gordijn 2012).

Here we will discuss our experiences incorporating bioethics in our classes.

What are we talking about when we talk about teaching bioethics?

When discussing the teaching of bioethics, an initial question arises: what kind of bioethics do we teach? There is no such thing as a unique theory of bioethics; in fact, there are *different bioethics*. To talk about how we teach bioethics, we first need to clarify what we mean by bioethics. Considerations about what bioethics is and how it relates to ethics, professional ethics, and moral philosophy, directly determine how it is taught.

The first educational trend in bioethics is based on moral philosophy. As Ferrater Mora (2001) pointed out, the term *ethics* derives from the Greek *ethos,* which means "habits"; therefore, ethics has often been considered a doctrine of habits. As the meaning of the word has evolved, *ethical* has become increasingly identified with *moral,* and *ethics* has come to mean the science that deals with moral objects in all their forms, in other words moral philosophy. In the philosophical tradition the term *moral* is used to describe value systems, and the term *ethics* to denote the discipline that studies moral objects.

This separation between ethics and morality has been preserved in teaching bioethics, which has generally focused on morality, following a historical tradition strongly impregnated with moral philosophy from Ancient Greece.

Plato (c. 427–347 A.C.) and Aristotle (384–322 A.C) constitute unavoidable references when considering ethical issues in classical Greece. The concern for the good life or good living gives ethics a central place in their inquiries. For Plato, ethical questions are associated with the constitutive desires of the human *psykhé* (as indicated in his *Banquet, Republic,* and *Phaedrus*), with the corporeal body as an obstacle or vehicle for access to wisdom (as in *Phaedon* and *Timaeus*), and with justice as a virtue that regulates the citizen's relationship with others (as in *Gorgias* and *Republic*). On the other hand, Aristotle reflects on the nature of humanity (in, among others, *De Anima, Politics,* and especially the *Nicomachean Ethics,* dedicated to his son Nicomachus, which presents a systematic formulation of ethics as a discipline, as well as the value of teaching ethics). Therefore, from its origins, ethics has been taught as a reflective way of thinking. Socratic reflection, described by Plato in the *Socratic dialogues,* involved a method of dialectics or logical demonstration in the investigation of or search for new ideas. The search for the truth thus implied reflection and understanding between two or more people, who reasoned together.

Some centuries later, in Modernity, we find two of the greatest exponents of ethics that continue to influence the discipline to this day: René Descartes (1596–1650) and Immanuel Kant (1724–1804). In the metaphor of the tree of knowledge that Descartes presents in his *Principles of Philosophy,* ethics is placed alongside mechanics and medicine, as one of the fruitful branches of the tree, of which metaphysics is the root and physics, the trunk. Ethics is seen as the science of how to enjoy a full and rewarding life. Descartes considers that living happily consists of having a perfectly contented and satisfied mind; the correct use of reason is the key to achieving this mental satisfaction, and passions are the greatest obstacle to achieving this rationally guided life. Thus, for Descartes, the fundamental use of wisdom is to teach us to dominate our passions and to keep the arbitrary will to guide the rationality of the intellect. The idea of rejecting emotions and passions has had a profound effect on our model of education.

On the other hand, Kant sought to establish a supreme principle of morality, a standard that would allow us to correctly judge any type of custom. This principle must have an absolute validity and must therefore be found a priori in concepts of pure reason, that is, in universal knowledge with the character of internal need, independent of experience. The principle of morality for Kant is closely related to duty. Thus, he formulates his categorical imperative, which compels one to act in such a way that one would want to become a universal law. This moral principle has also taken deep roots in our culture: it is reflected not only in our efforts to teach what it is right and wrong, but also in our thinking that there can be criteria of good and evil that serve all human beings equally. A single, unified imperative can be difficult to formulate in bioethics. One alternative is the concept of dignity (Foster 2011), although it can be underestimated (Pinker 2008). In fact, the term is also polymorphic, and it is difficult to apply when different cultural and historical values are involved.

Both Descartes and Kant have influenced the teaching of bioethics. From Modernity to the present, ethics has been considered nearly synonymous with morality. We also think about practical reason, differentiating it from pure reason or theoretical reason. One strong current in education and thought considers it imperative to eliminate the influence of emotions and feelings. Especially in medicine, this emphasis on rational thinking has shaped education for many years.

In the nineteenth century, utilitarian ethics become relevant to moral philosophy. Jeremy Bentham (1748–1832) and John Stuart Mill (1806–1873) pointed out that all human actions, norms, laws, or institutions must be judged according to their usefulness and the "greatest happiness principle". This new ethic is based on maximizing the enjoyment of life while minimizing sacrifice and suffering. The purpose of every human being is the pursuit of pleasure; in this scheme, utility refers to the pleasure (and the avoidance of suffering) that norms, actions, and institutions should produce in people. This utilitarian ethic gained strength with the technoscientific developments of the mid-19th and 20th centuries: if technoscientific advances exist, why not use them? Teaching bioethics from this point of view can carry the risk that "everything under the sun is permissible" without questioning the limits or taking each person's responsibility into account.

It is interesting to include here a contemporary definition of ethics. The French philosopher Alain Badiou (1937–) points out that it is essential to understand that ethics involves human practices, not only concerning morality and/or values, but also because of the subjective action that it represents (2002). Teaching bioethics from this point of view requires us to center on the person making decisions in the midst of a bioethical quandary.

Besides, bioethical conflicts are not defined solely by shared social values. Views of bioethics, morals, values, and legal aspects are always diverse: there is no worldwide cultural consensus about values. Cultural differences must be taken into account, and we should think about bioethics in the plural and teach various conceptualizations of bioethics.

Another necessary distinction in teaching bioethics is that bioethics overlaps with but is different from legality, that is, what is allowed by the laws in each country and what is not. First, it is necessary to distinguish between what is legal (according to the law) and what is legitimate (according to ethics): not all laws are ethical, and not all ethical acts are legal, although in an ideal world these two aspects would be identical. Bioethics must therefore dialogue with positive law, but it must avoid the temptation to fall into a legalism that confuses what is legal with what is ethical. Thus, knowledge does not precede the situation but is thought of as a whole: the confrontation of legal regulations with a concrete case requires the consideration and interpretation of the subject.

Teaching bioethics should not aim to make students behave in a singular or particular way: not all actions can be imitated. In bioethics, prescribing solutions can only lead to failure because there is no single answer for all cases. As Diego Gracia (2016) and Montserrat Esquerda (2019) pointed out, teaching bioethics is not only linked to knowledge and skills, but to the attitudes and people's character.

Moreover, teaching bioethics involves more than just teaching professional ethics or professionalism, in other words, what a professional should do in a specific case. Bioethics can also be taught with situations in the everyday lives of ordinary people.

In teaching bioethics, it may be necessary to distinguish between professional and personal ethics (Downie and Clarkeburn 2005). We need to ensure that students, especially health sciences students, acquire a thorough understanding of correct decisions in their professional practice; however, regarding personal ethics, our aims should be different, because individuals have different views on current issues like animal experimentation or human cloning. Therefore, we must focus on helping students develop clear thinking while exposing them to a range of differing views (Downie and Clarkeburn 2005).

Two main lines can be defined in contemporary medical ethics education, as well in bioethics: the first aims to develop "virtuous" doctors, promoting traditional values such as responsibility, honesty, and commitment; the second focuses on teaching practical skills to resolve ethical quandaries, providing future professionals with a series of tools to be able to detect and respond to ethical problems in professional practices (Eckles 2005).

This distinction is extremely important, not only for deciding on the contents of bioethics that should be taught but also for deciding how to teach autonomous thinking; specially taking into account that, as ten Have (2014) pointed out, in the neoliberal globalization of our context, bioethics should be a critical discourse that analyzes and question the current value systems.

The deliberation method

The traditional approach to teaching in which the teacher lectures the students is insufficient when the goal is to teach students to think for themselves. Lectures need to be supplemented or replaced with other methods. One transversal method that can be used in teaching bioethics is the deliberation method (Gracia 1989, 1991, 2000), which emphasizes rational analysis for decision making through Socratic reflection and sharing reflections with others.

Deliberation method can be considered as the best method to find the "best", "reasonable", "prudent" or "wise" solution (Gracia 2016). Its goal is not only to discuss the facts but is also to give reasons on that facts, like values and beliefs, which sometimes are not completely rational.

Deliberation method requires a series of steps, but rather than being mechanical or exact, these steps involve listening and dealing with conflict in the face of uncertainty.

The consecutive steps required in bioethical deliberation are:

1) *Deliberation about facts*. First, the case or the bioethical problem is presented. In the biomedical field, the patient's clinical history is often a crucial document. Later, these facts are analyzed, and the details of the case are specified and clarified.
2) *Deliberation about values*. First, the moral problems involved (i.e., those that refer to the sphere of values) must be identified. The recommended approach is to pose questions to help students identify these problems themselves. Second, the fundamental moral problem must be identified by ranking all the moral problems and choosing the most relevant one. Third, the conflicting values in the chosen problem must be identified.
3) *Deliberation about duties*. First, extreme courses of action must be identified. An example of an extreme course of action would be opting for one value in a conflict between two values to the total detriment of the other value. It is necessary to expose these extreme courses so they can be avoided. If we talk about two values in conflict, we need to

examine how both are important in the situation. Second, intermediate courses of action must be identified to bridge the two conflicting values. To do this, a dilemma (a word whose origin suggests only two possible solutions) must be turned into a problem (for which we can seek various complex solutions). Third, the optimal course of action must then be identified. This is the course that results in the least harm to conflicting values; the optimal course must take into account the circumstances of the case and the consequences of the decision. There may be many good courses, but students need to work toward finding the best one.

4) *Deliberation about final responsibilities*. First, safety tests must be done on the decision. This step seeks to ensure that the decision has not been rushed (test of time), that it can be argued publicly (test of publicity), and that it takes into account the legal aspects that may be involved (test of legality).
5) *Final decision*. This is the decision made by the person who introduced the case, who must decide on the best solution. The deliberation process is consultative. Ultimately, the responsibility always rests with the person in charge of the case.

About methods and resources: how narrative bioethics can help us focus on a different bioethics

Contributions of hermeneutical philosophy (e.g., from H.G. Gadamer, P. Ricoeur, and J. Ortega y Gasset) challenge the very idea of rationality and reveal the shortcomings of the rationalist approach forged by the philosophical tradition, focusing instead on the importance of narration in human experience.

Narrative bioethics supplements the rational approach that excludes the world of emotions and feelings that dominated ethics from the Ancient Greeks through Modernity. Narrative bioethics includes not only norms and principles, but also circumstances and emotional aspects. Domingo Moratalla and Feito Grande (2014) propose narrative bioethics as a new way of understanding the discipline and its practical dimension. This transformative approach reviews and criticizes the dominant model in bioethics:

1) Whereas the traditional model is based on individualistic autonomy, narrative bioethics proposes a relational autonomy in which the autonomous subject is understood not as an individual isolated from the world, but rather as a subject characterized by interrelation and interdependence with other people.
2) Instead of the prevailing model of general principles, narrative bioethics fosters a more contextual and situational ethics, claiming

that truth and knowledge are always situational and partial, and therefore no decontextualized absolute or universal truths can exist.

3) Unlike the traditional model in which rational and abstract argumentation and justification exclude feelings and emotions because they are considered labile and confused, narrative ethics defends an ethic of plural dialogue where responsibility is the most important component. Indeed, the analysis of bioethical issues from this perspective can also incorporate the study of the paradigm of responsibility, since it focuses on "what is happening" instead of "what we should do".

From this perspective, two aspects of narrative bioethics become clear. This approach defines humanity as a narrative species and narration as a fundamental human activity (Charon 2008). We use stories to live or decide better, and the most important source of knowledge of life, culture, and morals are narrative traditions. Narratives define views of the world. Stories, whether from novels, songs, folklore, movies, travel narratives, or casuistry, shape our actions, and this makes them very useful in teaching bioethics.

Medical case studies

One of the most widely used narrative methods for teaching bioethics is *Casebooks* or *Case Studies*. Frequently used in medicine, these "medical case stories" allow us to test personal ideas and legal frameworks by applying them to individual cases involving medical and moral dilemmas (Chambers 2001). Case studies can also be considered a particular type of story or literary genre that can be analyzed within narrative theory (Kermode 1967, 1979, Hauerwas and Burrel 1977).

The most widely used method for analyzing bioethical situations exploits clinical cases like those comprising UNESCO's Core Curriculum (2001), a set of "true story" narratives giving brief details of the bioethical dilemma involved. These narratives come from cases submitted by doctors from their practice after masking sensitive and confidential data.

As an example, let us briefly detail one of the cases presented by UNESCO about Informed Consent (UNESCO 2001:34).

The right to refuse treatment. Case Report No. 15

A 57-year-old man, affected by a throat cancer with widespread metastasis, at the last stage, is hospitalized. The medical staff realizes the serious condition of the patient and that his lucidity from time to time seems compromised. The medical team feels he may need intubation to support his life functions and breathe better. They ask the patient about the intubation in the morning and he agrees. During the afternoon, when the man is conscious, he seems unsure about his previous agreement, refusing the intubation support. The next day the situation repeats itself. Dr. Francesco Masedu and Prof. Ferdinando di Orio. Italy

Reflections. To intubate or not to intubate, that is the question

Select an option:

1) Intubate, based on the principle of beneficence in the clear and continuous absence of stated opposition by the patient.
2) Do not intubate, because the last stated preference of the patient is to refuse the intubation support.
3) Intubate in the morning.
4) Do not intubate now, nor intubate in case of life-threatening obstructive respiratory failure. (The person would die in this case.)
5) Do not intubate now, but (try and) intubate in emergency if a severe respiratory failure threatening the patient's life functions occurs.

The patient is under no obligation to stay healthy or to receive any kind of treatment. He has the right to refuse or to stop a medical intervention. He is free to choose treatment or no treatment at all, or just partial treatment.

Although five possible solutions for this situation are stated, the decision is linked to the dilemma: to intubate or not to intubate. There are no data on the patient's history, his relatives, or previous treatment. The descriptive elements are limited in quantity and depth.

Solbakk (2006) synthesizes this model, taking into account the following aspects: (1) the stories are said to be authentic (i.e., they deal with events that have actually occurred); (2) it is stated that they are direct and brief, and that therefore they save time since the dilemmas involved emerge easily when proposing two possible actions available to answer the case; and (3) they are supposed to be very accessible, as they promote rapid comprehension and understanding among students in an inexpensive way that does not require too many resources.

In this regard, he points out that these are "apparent didactic advantages". The criticism enunciated by Solbakk (2006) can be summarized as follows:

1) These stories are usually spread in their most anemic form: they never give a complete vision of what happened since they only constitute condensed versions of the story or original event.
2) The version that we know of the stories has generally been constructed by whoever holds the power (in most cases, the medical doctor).

3) The "selective" nature of the stories based on real cases conveys the message that, in situations where the conflict is bioethical, the possibility of choosing ends up being reduced to practically closed options.
4) Because these stories primarily focus on dilemmas, true case histories tend to reduce the problem to conflict resolution in moral terms.

Although case stories may very well have these drawbacks, it can be very interesting to organize classroom debates about them (and in courses for health sciences students, they may be their first contact with bioethics in the context of clinical experience).

Many examples of casebooks can be downloaded for free from UNESCO (http://www.int-chair-bioethics.org/?page_id=41), including the following documents: Informed Consent; Reproductive Health; Ethics in Psychiatry and Mental Health; Bioethics for Judges; Bioethics and the Holocaust; Teaching Ethics in Organ Transplantation and Tissue Donation Cases and Movies; and Psychiatric Ethics and the Rights of Persons with Mental Disabilities in Institutions and the Community, among others.

Literature

Literature can be a splendid resource to help students think about different perspectives in bioethics (Loscos et al. 2008, Lancet 2015, Torrens et al. 2018), although its role has been deeply misunderstood.

Chambers (2016) points out that rather than simply portraying the world as it really is, narrative fiction forcefully presents a specific perspective of the world.

Literature can help us to better understand life and sickness and to see the power and implications of clinical practice as well as the effects of disease on the patient. By showing us patients' experiences and lives, it can help us develop empathy and broaden our perspectives to enrich our approach to resolving bioethical conflicts (Baños and Guardiola 2018).

Research has shown that students learn best when what they are learning elicits a positive emotional response (Taylor 2010, McConnell and Eva 2012), and literature can engage students with complex stories and characters. Emotional engagement enables students to understand the situation better (Lake et al. 2015), although reflection is also an essential component in learning this way (McCann and Huntley-Moore 2016).

Baños and Guardiola (2018) introduce an important question: why should medicine be interested in literature?

"Firstly, it is important to remember that physicians were traditionally educated individuals within the society. In fact, they were the first professionals to be educated in Western universities, starting from the thirteenth century. Secondly, clinical experience is often too intense to be understood and accepted by following only biological patterns. Finally, a narrative can include nuances that are absent from the traditional medical discourse" (p. 215).

Many goals in teaching bioethics can be achieved through literature (Squier 2000): to deepen students' knowledge of both patients' and doctors' expectations in their understanding of the doctor-patient relationship, to prepare and motivate students during their training in the techniques necessary for medical consultations, to prepare students for their years of training in clinical practice by improving their understanding of the psychosocial aspects of disease and helping them to behave empathetically towards their patients, to develop a deeper understanding of how human beings communicate, and to stimulate self-reflection and moral imagination.

We must add that literature can also warn us about future threats to the doctor-patient relationship and to humanity. Non-fiction and fiction, including short stories, novels, and even poems, can serve this purpose. Some rich examples are classic novels like Mary Shelley's *Frankenstein,* George Orwell's *1984,* Lewis Carroll's *Through the Looking-Glass, and What Alice Found There,* Ian McEwan's *The Children Act,* and José Saramago's *Blindness* and *Death with Interruptions.*

Feature films and TV series: the usefulness of cinemeducation

Students are generally very interested in popular movies and TV series. Both can help teachers introduce students to complex scenarios that can be difficult to understand through traditional educational methods. These materials are a very valuable but inexpensive resource that promotes learning through emotional engagement (Coon 2018).

Like case stories and literature, scenes from films and TV series can be used to engage students and to introduce elements for discussion. They can also be used to identify students' preconceptions, as well as to help them develop a critical spirit and a scientific attitude. The potential usefulness of these media in teaching has been explored in different ways, and different ways of using films in the classroom and different hypotheses about the pedagogical efficacy of audiovisual material have been tested.

Cinemeducation is a rigorous methodology based on the application of entire films, fragments of films, or television series for specific purposes in medical education (Alexander et al. 1994, Alexander 2002, Alexander et al. 2005, Alexander 2012). This approach can be useful for introducing students to complex scenarios, and not only strictly medical ones (Cambra Badii and Baños 2020).

In *cinemeducation,* films can constitute the main material of the class and are not just an entertaining component to complement other materials. Our proposal for working with films (Cambra Badii and Baños 2020) considers adopting this educational activity as a scientific investigation, proposing basic hypotheses (usually to test the efficacy of specific cinematographic materials for teaching a particular health sciences subject), and testing them during class. It is necessary to build a rigorous protocol for the entire sequence of the activity, starting with choosing the audiovisual materials (an entire film or TV series episode or one or more clips from these productions), then designing the teaching activity, and finally assessing the effectiveness of the film-based activity (Cambra Badii and Baños 2020).

Here are some films that we find useful for different teaching purposes:

- Helping students understand the experience of the disease: *The Elephant Man* (David Lynch 1980), *My Left Foot* (Jim Sheridan 1989), *Awakenings* (Penny Marshall 1990), *Lorenzo's Oil* (George Miller 1992), *Elling* (Petter Næss 2001), *Big Fish* (Tim Burton 2003), *The Painted Veil* (John Curran 2006), and *Dallas Buyers Club* (Jean-Marc Vallée 2013).
- Exploring the doctor-patient relationship and learning the importance of communication: *Not as a Stranger* (Stanley Kramer 1955), *The Doctor* (Randa Haines 1991), *Patch Adams* (Tom Shadyac 1998), and *Hippocrates* (Thomas Lilti 2014).
- Discussing the beginning of life, particularly concerning abortion: *If these Walls Could Talk* (Cher and Nancy Savoca 1996), *The Cider House Rules* (Lasse Hallström 1999), *The Secret of Vera Drake* (Mike Leigh 2004), *4 Months, 3 Weeks and 2 Days* (Cristian Mungiu 2007), *Mother and child* (Rodrigo Garcia 2009), and *La Innocència* (Lucia Alemany 2019).
- Discussing the end of life, particularly concerning euthanasia, assisted suicide, and palliative sedation: *The Barbarian Invasions* (Denys Arcand 2003), *Million Dollar Baby* (Clint Eastwood 2004), *The Sea Inside* (Alejandro Amenábar 2004), and *The Wings of Life* (Antoni P. Canet 2006).
- Understanding ethical issues in research: *Awakenings* (Penny Marshall 1990), *Extreme measures* (Michael Apted 1996), *The Tuskegee Experiment* (Joseph Sargent 1997), *Gattaca* (Andrew Niccol 1997), *Something the Lord Made* (Joseph Sargent 2004), and *The Constant Gardener* (Fernando Meirelles 2005).
- Appreciating the impact of disasters and catastrophes: *Alive!* (Frank Marshall 1993), *The Pianist* (Roman Polanski 2002), *Ice Age* (Chris Wedge 2002), *The Impossible* (Bayona 2012), *Sully* (Clint Eastwood 2016), *Don't Look up* (Adam McKay 2021).
- Understanding pandemics: *12 Monkeys* (Terry Gilliam 1995), *Children of Men* (Alfonso Cuarón 2006), *Blindness* (Fernando Meirelles 2008),

Contagion (Steven Soderbergh 2011), and *Train to Busan* (Yeon Sang-ho 2016).

- Discussing animal rights: *Gorillas in the Mist* (Michael Apted 1988), *Babe* (Chris Noonan 1995), *Ojka* (Bong Joon-ho 2017).

TV series have a special place in *cinemeducation*. In the last two decades, they have become the dominant audiovisual format. Every day, millions of people watch TV series from different genres ranging in format from the traditional sitcom to miniseries comprising only a few episodes with similarities to feature films. Online platforms such as Netflix or Amazon have eliminated time constraints on viewing, providing full immediate access to seasons on televisions, computers, tablets, and cell phones, and thus allowing viewers to become fully immersed in a series at their own pace. The wide availability of TV series has made them an ideal gateway for motion-picture fiction.

Many medical and nursing students follow medical dramas (Czarny et al. 2008, Weaver and Wilson 2011, Weaver et al. 2014, Williams et al. 2014, Cambra Badii et al. 2021). In the last decade, investigations into the use of these series in educating health sciences students, especially for analyzing and teaching bioethical issues, have shown that they are a rich resource. For example, Czarny et al.'s (2010) content analysis of 50 episodes of *Grey's Anatomy* (Rhymes 2005–) and *House MD* (Shore 2004–2012) found 79 depictions of bioethical issues from the 2005–2006 television season, classifying them under 11 topics, including consent, ethically questionable departures from standard practice, death and dying, and confidentiality. Cambra Badii et al.'s (2020) content analysis of the first season of *The Good Doctor* (Shore 2017–) found 179 situations that can be used to teach bioethics. Arawi (2010) proposed working with vignettes from *Grey's Anatomy* and *House MD* to teach biomedical ethics, and Pavlov and Dalquist (2010) used *Grey's Anatomy* to teach communication and professionalism. In an earlier study, Spike (2008) pointed out bioethical issues involved in doctor-patient communication and medical errors depicted in *Scrubs* (Lawrence 2001–2010).

Recently, medical dramas depicted the coronavirus pandemic, and vignettes of *The Good Doctor* (Shore 2017–), *The Resident* (Holden Jones, Schore and Sethi 2018–), *Grey's Anatomy* (Rhymes 2005–), *Chicago Med* (Brandt, Haas and Olmstead 2015–) and *New Amsterdam* (2018–2023) have been used to teach bioethics in undergraduate classes (Cambra Badii et al. 2022).

Some tips for teaching strategies and assessment

In undergraduate students, bioethics can be approached as a specific subject within the curriculum or as transversal or integrated models where bioethics is incorporated in the teaching of traditional subjects. We propose mixing the two approaches to fulfill the traditional requirements for each course, while pointing out and discussing bioethical problems throughout the degree program. Regardless of whether students are pursuing degrees in the health sciences or fields that have no contact with patients, the same indications should apply, given that bioethics affects us all and anyone can be faced with bioethical problems.

In both approaches, teaching in small groups is highly recommended. An active dialogue among students and between students and teachers is necessary to ensure that the class can transmit critical knowledge, that interdisciplinary issues can be integrated, and that all students can express their opinions. However, teaching in small groups carries the risk of leaving out important topics and points of view, so rigorous methods and strict surveillance are essential.

Deliberations about bioethical situations must be firmly connected with real life. There is a danger of introducing a bias when constructing a fictitious case, so it is important to ensure that it does not only represent the teacher's position or the predominant cultural morality, thus separating bioethics from students' experience. Likewise, there is the risk of leading students to think that there is only one right option (often identified in the teacher's expectations), in which case bioethics analyses are reduced to a single option based on morality.

In secondary schools, we highly recommend working with transversal or integrated models, incorporating bioethics into nearly all subjects, including the sciences, literature, and even language studies. In this way, students can come to appreciate the role of bioethics in our lives without thinking of it as a subject isolated from others.

In university health sciences studies, students learn fundamental concepts during their early years that they will apply in practical training in clinical internships. In this context, it is essential to include bioethics as a specific subject to ensure that it is explicitly taught in the curriculum. It might seem tempting to wait until students face clinical dilemmas in their practical training, when they can identify with their teachers' ethical attitudes in their professional performance, but it is best to include education in bioethics from the first year to help students develop their sensitivity toward dilemmas that will arise in practice.

The insistence on using small groups in both university and secondary school must be counterbalanced by the need to ensure that students gain a solid grounding in bioethical principles because, although problem solving is a useful ability, it can be difficult to acquire these principles in fast-paced group interaction. The best way to teach bioethics is to combine theory with practical activities, even forgoing lectures (Mattick et al. 2006). Nevertheless, assessments should focus on practical applications rather than theoretical concepts. One of the most common ways of assessing students' development is to assign a short dissertation or deliberation on a specific topic (Stern et al. 2008). Topics can be drawn from resources such as case studies, literature, films, or TV series.

Conclusions: towards complex thinking

Traditionally, professors have transmitted their knowledge and imparted their doctrine as truth. However, knowledge is ever-changing, and there can be no single truth. This is especially relevant in dealing with bioethical problems. A plurality of voices, perspectives, and contents are essential in teaching bioethics.

Today, more than ever, technoscientific advances form part of our lives. Bioethics is a critical instrument for the analysis of scientific activity, allowing us to discuss its benefits in conjunction with the possible undesirable consequences it entails. For these reasons, bioethics should be integrated into the curricula at all levels of schooling. Moreover, bioethics should permeate fields far beyond the health sciences. At universities, bioethics should be taught throughout the trajectory and must encompass a wide range of subjects. This strategy aims to develop a responsible citizenry that is capable of making decisions in a rational and well-founded manner.

In its origins, bioethics was entwined with philosophy, and the strong influence of philosophy established a tradition of teaching bioethics in the same way that the humanities have been traditionally taught, mainly through lectures. However, more recent experiences have shown that a practical approach to teaching bioethics is very effective. Small groups allow all students to participate actively through dialogue and deliberation and help them to develop critical thinking and decision-making skills.

Bioethics cannot remain divorced from emotions. We need to be able to tolerate anguish and uncertainty related to our rapidly changing environment. The bioethical deliberation method promises to help us face the challenges posed by these changes, and clinical cases, literature, cinema, and series are extremely valuable resources that can be used with this method. A combination of different pedagogical methods may be more effective in helping students to acquire important theoretical knowledge and may improve decision-making ability. There is a strong consensus that teachers need to be up to date in both bioethical theory and teaching resources and that small classes are best for many activities.

In bioethics, theoretical knowledge is essential but insufficient. To enable the best decisions, students need to develop practical skills in resolving bioethical problems. It is crucial that students acquire theoretical bioethical knowledge and the ability to apply it in solving practical situations that they will encounter in their lives.

References cited

Alexander, M., M.N. Hall and Y.J. Pettice, 1994. Cinemeducation: an innovative approach to teaching psychosocial medical care. Fam Med. 26: 430–433.

Alexander, M. 2002. The Doctor: a seminal video for cinemeducation. Fam Med. 34: 92–94.

Alexander, M., P. Lenahan and A. Pavlov. 2005. Cinemeducation: A Comprehensive Guide to Using Film in Medical Education. Radcliffe Publishing, Oxford.

Alexander, M. 2012. Let's look at the data: a review of the literature. pp. 3–9. *In*: Alexander, M., P. Lenahan and N. Pavlov (eds.). Cinemeducation: Using Films and Other Visual Media in Graduate and Medical Education, 2. Radcliffe Publishing, London, UK.

Annas, J. 1999. Platonic Ethics, Old and New. Cornwell University Press, London.

Arawi, T. 2010. Using medical drama to teach biomedical ethics to medical students. Med. Teach. 32: 205–10.

Badiou, A. 2002. Ethics: An Essay on the Understanding of Evil. Verso, London.

Baños, J.E. and E. Guardiola. 2018. Literature in medical teaching: the crucial importance of literature in the education of medical students. Mètode. 8: 215–221.

Beever, J., S.M. Kuebler and J. Collings. 2021. Where is ethics taught: An institutional epidemiology. Int. J. Ethics Educ. 6: 215–238.

Cambra Badii, I. and J.E. Baños. 2020. The University goes to the movies: our experience using feature films and tv series in teaching health sciences students. pp. 105–148. *In*: Kim, S. (ed.). Medical Schools: Past, Present and Future Perspectives. Nova Publishers, New York, USA.

Cambra-Badii, I., A. Pinar and J.E. Baños. 2020. The Good Doctor and bioethical principles: a content analysis. Educ. Med. 22: 84–88.

Cambra Badii, I., E. Moyano, I. Ortega, J.E. Baños and M. Sentí. 2021. TV medical dramas: health sciences students' viewing habits and potential for teaching issues related to bioethics and professionalism. BMC Med. Educ. 21: 509.

Cambra-Badii, I., E. Guardiola and J.E. Baños. 2022. The COVID-19 Pandemic in Serial Medical Dramas. JAMA 327: 20–22.

Chambers, T.S. 2001. The fiction of Bioethics: a précis. AJOB 1: 40–43.

Chambers, T. 2016. Eating one's friends: fiction as argument in bioethics. Lit. Med. 34: 79–105.

Charon R. 2008. Narrative Medicine: Honoring The Stories of Illness. Oxford University Press, New York.

Coon, R. 2018. Cinema in nursing education: tapping into the affective domain. J. Nurs. Educ. 57: 188–189.

Czarny, M., R. Faden, M. Nolan, E. Bodensiek and J. Sugarman. 2008. Medical and nursing students' television viewing habits: potential implications for bioethics. Am. J. Bioeth. 8: 1–8.

Czarny, M., R. Faden and J. Sugarman. 2010. Bioethics and professionalism in popular television medical dramas. J. Med. Ethics 36: 203–6.

Domingo Moratalla, T. 2011. Bioética y Cine. De la narración a la deliberación. San Pablo, Madrid.

Domingo Moratalla, T. and L. Feito Grande. 2014. Bioética narrativa. Escolar y Mayo, Madrid.

Downie, R. and H. Clarkeburn. 2005. Approaches to the teaching of bioethics and professional ethics in undergraduate courses. Biosci. Educ. 5: 1–9.

Eckles, R.E, E.M. Meslin, M. Gaffney and P.R. Helft. 2005. Medical ethics education: where are we? Where should we be going? A review. Acad. Med. 80: 1143–1152.

[Editorial]. 2015. Literature and medicine: why do we care? Lancet. 385(9963): 90.

Esquerda, M., J. Pifarré, H. Roig, E. Busquets, O. Yuguero and J. Viñas. 2019. Evaluando la enseñanza de la bioética: formando «médicos virtuosos» o solamente médicos con habilidades éticas prácticas. Atención Primaria 51: 99–104.

Ferrater Mora, F. 2001. Diccionario de Filosofía. Ariel, Barcelona.

Foster, C. 2011. Human Dignity in Bioethics and Law. Bloomsbury Publishing, London.

Gracia, D. 1989. Fundamentos de bioética. Eudema, Madrid.

Gracia, D. 1991. Procedimientos de decisión en ética clínica. Eudema, Madrid.

Gracia, D. 2000. Fundamentación y enseñanza de la bioética. El Búho, Bogotá.

Gracia, D. 2016. The mission of ethics teaching for the future. Int. J. Ethics Educ. 1: 7–13.

Hauerwas, S. and B. Burrel. 1977. From System to Story: an alternative pattern for rationality in ethics. pp. 111–152. *In*: Engelhardt, H.T. and D. Callahan (eds.). The Foundations of Ethics and its Relationship to Science: Knowledge, Value and Belief. Hastings Center, New York, USA.

Kermode, F. 1967. The Sense of Ending: Studies in the Theory of Fiction. UOP, New York.

Kermode, F. 1979. The Genesis of Secrecy: On the Interpretation of Narrative. HUP, Cambridge.

Lake, J., L. Jackson and C. Hardman. 2015. A fresh perspective on medical education: the lens of the arts. Med. Educ. 49: 759–72.

Loscos, J., J.E. Baños, F. Loscos and J. de la Cámara. 2008. Medicine, cinema and literature: a teaching experience at the Autonomous University of Barcelona. JMM 2: 138–142.

Mattick, K. and J. Bligh. 2006. Teaching and assessing medical ethics: where are we now? J. Med. Ethics 32: 182–285.

McCann, E. and S. Huntley-Moore. 2016. Madness in the movies: An evaluation of the use of cinema to explore mental health issues in nurse education. Nurse Educ. Pract. 21: 37–43.

McConnell, M.M. and K.W Eva. 2012. The role of emotion in the learning and transfer of clinical skills and knowledge. Acad. Med. 87: 1316–1322.

Miles, A. 2009. On a medicine of the whole person: away from scientistic reductionism and towards the embrace of the complex in clinical practice. J. Eval. Clin. Pract. 15: 941–9.

Nussbaum, M. 2016. Not for Profit: Why Democracy Needs the Humanities. Princeton University Press, New Jersey.

Ordine, N. 2017. The Usefulness of the Useless. Paul Dry Books, Philadelphia.

Orefice, C., J. Pérez and J.E. Baños. 2019. The presence of humanities in the curricula of medical students in Italy and Spain. Med. Educ. 20: 79–86.

Osler, W. 1899. Address to the students of Albany Medical College. Albany Medical Annals 20: 307–309.

Pavlov, A. and G.E. Dahlquist. 2010. Teaching communication and professionalism using a popular medical drama. Fam. Med. 43: 25–7.

Pinker, S. 2008. The stupidity of dignity. The New Republic 28: 28–31.

Solbakk, J. 2006. Catharsis and moral therapy II: an Aristotelian Account. Medicine, Health Care and Philosophy 9: 141–153.

Solomon, M. 2015. Making Medical Knowledge. Oxford University Press, Oxford.

Spike, J. 2008. Television viewing and ethical reasoning: Why watching scrubs does a better job than most bioethics classes. Am. J. Bioeth. 8: 11–3.

Squier, H.A. 2000. Teaching humanities in the undergraduate medical curriculum. pp. 128–139. *In*: Greenhalgh, T. and B. Hurwitz (eds.). Narrative Based Medicine. BMJ Books, London, UK.

Stern, D.T., J.J. Cohen, A. Bruder, B. Packer and A. Sole. 2008. Teaching humanism. Perspec. Biol. Med. 51: 495–507.

Taylor, J.S. 2010. Learning with emotion: a powerful and effective pedagogical technique. Acad. Med. 85: 1110.

ten Have, H. and B. Gordijn. 2012. Broadening education in bioethics. Med. Health Care and Philos. 15: 99–101.

ten Have, H. 2014. Bioethics Education in a Global Perspective: Challenges in Global Bioethics. Springer, Pittsburgh.

ten Have, H. 2021. Ethics teaching as an infectious activity. Int. J. Ethics Educ. 6: 213–214.

Torrens, M., M. Farré, F. Fonseca and J.E. Baños. 2018. Literature as a teaching tool for medical students. pp. 60–67. *In*: Orefice, C. and J.E. Baños (eds.). The Role of Humanities in the Teaching of Medical Students. Monographs of the Esteve Foundation 38. Barcelona, Spain.

[UNESCO] Core Curriculum about Informed Consent. 2001.

Weaver, R.I. and Wilson. 2011. Australian medical students' perceptions of professionalism and ethics in medical television programs. BMC Med. Educ. 11: 50.

Weaver, R.I. Wilson and V. Langendyk. 2014. Medical professionalism on television: student perceptions and pedagogical implications. Health 18: 597–612.

Williams, D., D. Re and G. Ozakinci. 2014. Television viewing habits of preclinical UK medical undergraduates: further potential implications for bioethics. AJOB Empir Bioeth. 5: 55–67.

Index

About the Editors

Irene Cambra-Badii

Bachelor degree in Psychology from Universidad de Buenos Aires and PhD in Psychology from Universidad El Salvador, Buenos Aires, Argentina. Professor of Ethics at Universidad de Buenos Aires (2008–2018). Professor of Bioethics (Medical school and Health sciences studies) at the Universitat de Vic – Universitat Central de Catalunya (UVic-UCC), Spain, since 2021. Researcher at the Chair in Bioethics and Chair in Medical Education of UVic-UCC, since 2019. Member of the Research Group on Health Sciences Education in the Medicine and Life Sciences Department (Universitat Pompeu Fabra) since 2019.

Ester Busquets-Alibés

Registered Nurse from Universitat de Vic and PhD in Philosophy from Universitat Ramon Llull (URL). Professor of Ethics and Bioethics at Universitat de Vic – Universitat Central de Catalunya since 1999. Collaborator of the Borja Institute of Bioethics from URL (2001–2017). Director of de Journal Bioethics & Debat (2008–2017). Coordinator and researcher at the Chair in Bioethics of Universitat de Vic – Universitat Central de Catalunya since 2018. Member of different ethical committees (healthcare and research).

Núria Terribas-Sala

Degree in Law from Universitat de Barcelona. Specialized in bioethics and biolaw since 1992. Director of the Borja Institute of Bioethics -URL (1999–2014) and Director of de Journal Bioethics & Debat (2000–2017). Current Director of Victor Grifols i Lucas Foundation (2014-) and Director of the Chair in Bioethics at UVic-UCC (2015-). Vice-President of the Bioethics Committee of Catalonia, member since 2009. Member of the Commission for the guarantee and evaluation of euthanasia in Catalonia (2021-). Member for the Integrity of Research of Catalonia (2021-). Member of the Advisory Council on Health – Department of Health of the Government of Catalonia (2018-); Member and legal advisor to clinical ethics committees.

Josep-E. Baños

MD and PhD in Pharmacology from Universitat Autònoma de Barcelona. Vice-dean and vice-chancellor of Teaching and Academic Affairs at Universitat Pompeu Fabra. Currently full professor of Pharmacology and Chancellor of Universitat de Vic – Universitat Central de Catalunya (since 2019). Executive Secretary of Spanish Society of Pharmacology. Member of ethical committees at Taulí Hospital (1993–2000), Universitat Autònoma de Barcelona (2000–2002) and Barraquer Foundation (2004–2019).

For Product Safety Concerns and Information please contact our EU representative GPSR@taylorandfrancis.com
Taylor & Francis Verlag GmbH, Kaufingerstraße 24, 80331 München, Germany

www.ingramcontent.com/pod-product-compliance
Ingram Content Group UK Ltd.
Pitfield, Milton Keynes, MK11 3LW, UK
UKHW020202200926
475702UK00004B/574
* 9 7 8 1 0 3 2 2 1 7 5 5 0 *